-D'S COL
WITHDRAWN FROM
THE LIBRARY
UNIVERSITY OF
WINCHESTER
KA 0035948 3
D0655510

Athletic Ability & the Anatomy of Motion

Rolf Wirhed

Translated by
A.M. Hermansson

Wolfe Medical Publications Ltd

Published by Wolfe Medical Publications Ltd, 1984
Printed by Butler & Tanner Ltd, Frome, England
ISBN 0 7234 0854 8

Originally published in Sweden, under the title
'Anatomi och rörelselära inom idrotten' 1982
(Samspråk Förlagsaktiebolag)

For a full list of other titles published
by Wolfe Medical Publications Ltd, please
write to the publishers at: Wolfe House,
3 Conway Street, London W1P 6HE, England

CONTENTS

PREFACE

I have taught the subject of Kinetics (or the Theory of Motion) at the Örebro College of Physical Education (GIH) for a number of years. This subject deals with the mechanical laws that govern the technical execution of various exercises. During recent years, sportsmen have become increasingly interested in analysing the different aspects of, above all, technical kinds of sports (athletics, gymnastics, racket sports. . .). For example, instructors who train people in keeping fit are becoming more aware of the need to teach correct movements in order to increase the effect of training and reduce the risk of injury. Exercises designed to prevent injury are now an important feature of almost every sportsman's general training.

My own concern for these things was aroused by my keen interest in sports and the training I received in atomic physics which required, among other things, a knowledge of mechanics. I also studied anatomy in Uppsala so that I would better understand the mechanisms that govern the human body when it performs various movements.

This book takes up ideas that are usually summarized under the heading "Biomechanics". The lack of suitable literature has resulted in a dearth of anatomical and mechanical knowledge among sports leaders and the majority of them have felt the need for suitable literature and special sports associations have clearly expressed their desire that this need be met.

It is my hope that the readers of this book (including sports leaders, physical education teachers, physiotherapists) will find that examples contained in and ideas taken from this book will be of substantial practical value. The book may be used as a textbook for prospective physical education teachers, sports leaders and youth recreation leaders, and it may form the base for adult education classes. Students of the natural sciences at the upper secondary school level may find the book interesting as extra reading or may be inspired by it when working on special projects. It is my hope that after having read this book, you will have acquired a sound knowledge of the position and function of muscles, forces, the movement of force, the centre of gravity and the moment of inertia, and that you will find it easier to judge how and what to train. Hopefully, the general knowledge of anatomy and mechanics contained in this book will help you to gain new ability to analyze and understand the kind of sport you are interested in.

Rolf Wirhed
University Lecturer in Biomechanics

INTRODUCTION

This book deals firstly with general characteristics of bones, joints and muscles and it provides general guidelines for flexibility and strength training. Thereafter, the anatomy and function of the different parts of the body are described.

The main stress has been laid upon the analysis of movement, i.e. the biomechanical aspects of the exercises sportsmen perform to train a special part of the body. In the last section, such different types of movements as take off, rotation, flight, landing, etc., are dealt with. In the cases where the latin name is usual, it is given in brackets.

If you read the book on your own, you should always try to perform the movements or exercises. Feel the muscles that are described. Even if the book is read in a class, the exercises and movements should be carried out, at least to the extent that the situation allows. Anatomy and kinetics should be experienced in practice — they are not solely theoretical concepts.

Get to know your body! You have only got one and it must last your entire life.

Chapter 1

GENERAL ANATOMY OF BONES, JOINTS, AND MUSCLES

A. The Skeleton

The aim of this general section on the bones of the human body is to provide the reader with the basic knowledge needed to benefit from the literature on sports injuries, and also to aid him in analysing how the bones react to the stresses and strains to which they are subjected in different training exercises.

The skeleton of the body is made up of a variety of bones which are classified as follows:

short bones (e.g. bones of the ankle and wrist),
long bones (e.g. bones of the hand and forearm, thigh bone) and
flat bones (e.g. bones of the skull, breastbone).

The process of bone formation is called ossification. The flat bones (e.g. the bones of the skull) are developed in one stage from connective tissue. This developmental process is called direct ossification (intramembranous ossification). The bones of the skull are not fully formed in the newborn baby. The areas of incomplete ossification, the so-called fontanelles, can be felt with the fingers. Most of the bones of the skeleton are formed by indirect ossification (intracartilaginous or endochondral ossification). A cartilaginous model of the future bone is developed in the embryo and is later dissolved and replaced by bone.

Short bones are formed by indirect ossification. The cells at the centre of the growing cartilaginous model die. The so-called osteoblasts (os = bone, blast = immature cell) migrate from the membrane which surrounds the cartilage (periosteum) to the spaces left by the wasted cells. These osteoblasts gradually transform into bone cells (osteocytes). Not all the cartilage ossifies, some parts of it remain in the form of articular cartilage.

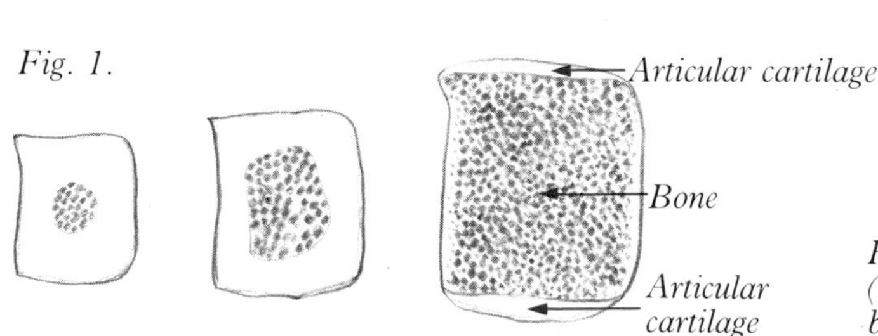

Fig. 1.

Articular fluid — which is found in the cavity of synovial joints — provides the articular cartilage with nourishment. The blood vessels which lie inside the bone tissue beneath the plates of articular cartilage also supply nourishment to the cartilage. The nutrients diffuse through the cartilage cells (being pressed or massaged through) and are not transported through small blood vessels (capillaries). Thus, if the joints are subjected to suitable all-around stress during the formative years of childhood, they will be well supplied with nourishment. This in turn leads to a thickening of the cartilage, thereby providing better protection against the kind of injury caused by excessive stress. A joint which is not subjected to any kind of stress, or is not fully used in all possible directions, reacts in an opposite manner; that is, the cartilage thins out.

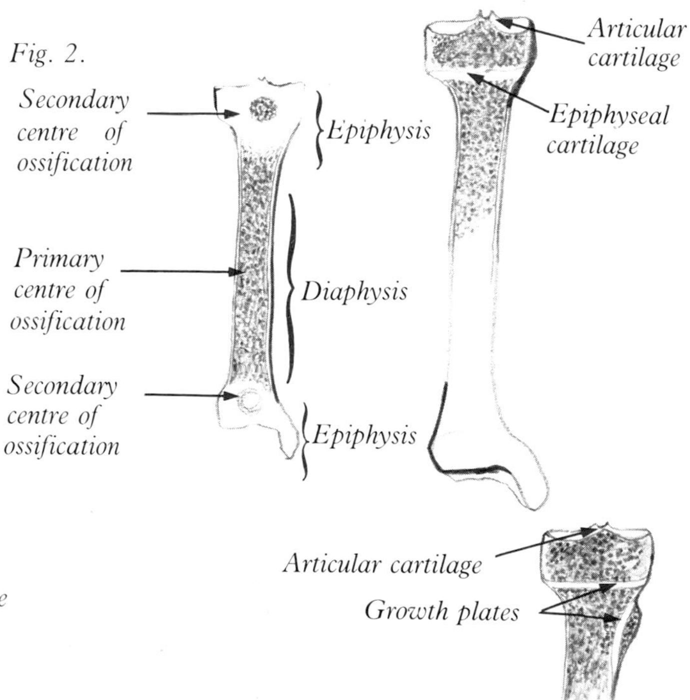

Fig. 2.

Fig. 2. This shows the growth plates situated at the proximal (towards the centre of the body) end of the shin bone shaft just beneath the insertion of the knee extensor muscles.

The long bones are also formed by indirect ossification, but here several so-called ossification centres can be found. In the formation of these bones, articular cartilage and growth plates (epiphyseal cartilage) are left between the shaft (diaphysis) and each extremity (epiphysis). The epiphyseal cartilage usually ossifies in the late teens or when a person is fully-grown. The so-called epiphyseal cartilage can be located in a youth with the aid of an X-ray.

Pathological changes at the growth plates are often due to hormonal disturbances. They can also be a consequence of incorrectly loading or overloading the skeleton. Compared with the rest of the body the skeleton grows very quickly during the first few years of life and during adolescence. Athletically active young people should therefore refrain from *extreme* strength training during puberty. A good recommendation is that the child, while still undergoing the changes of puberty, use only his or her own weight as a load when training. Thus, only the postpubertal child should include weights and equipment in his or her strength training routines.

A rather common complaint among children between the ages of 10 and 16 years is Osgood-Schlatter's disease. This condition results from excessive tension in the knee extensor's insertion on the small prominence of the shin bone (tibial tuberosity). The growth plates are irritated and the ensuing growth may be accelerated. The enlargement can be seen with the naked eye when the affected leg is compared with the healthy one. The afflicted child may have difficulty in kneeling on hard surfaces.

Figure 3 shows the fully-grown bone in detail. Microscopically small bone cells (osteocytes) lie embedded in a tissue consisting of collagen fibres which are highly resistant to force, inorganic salts

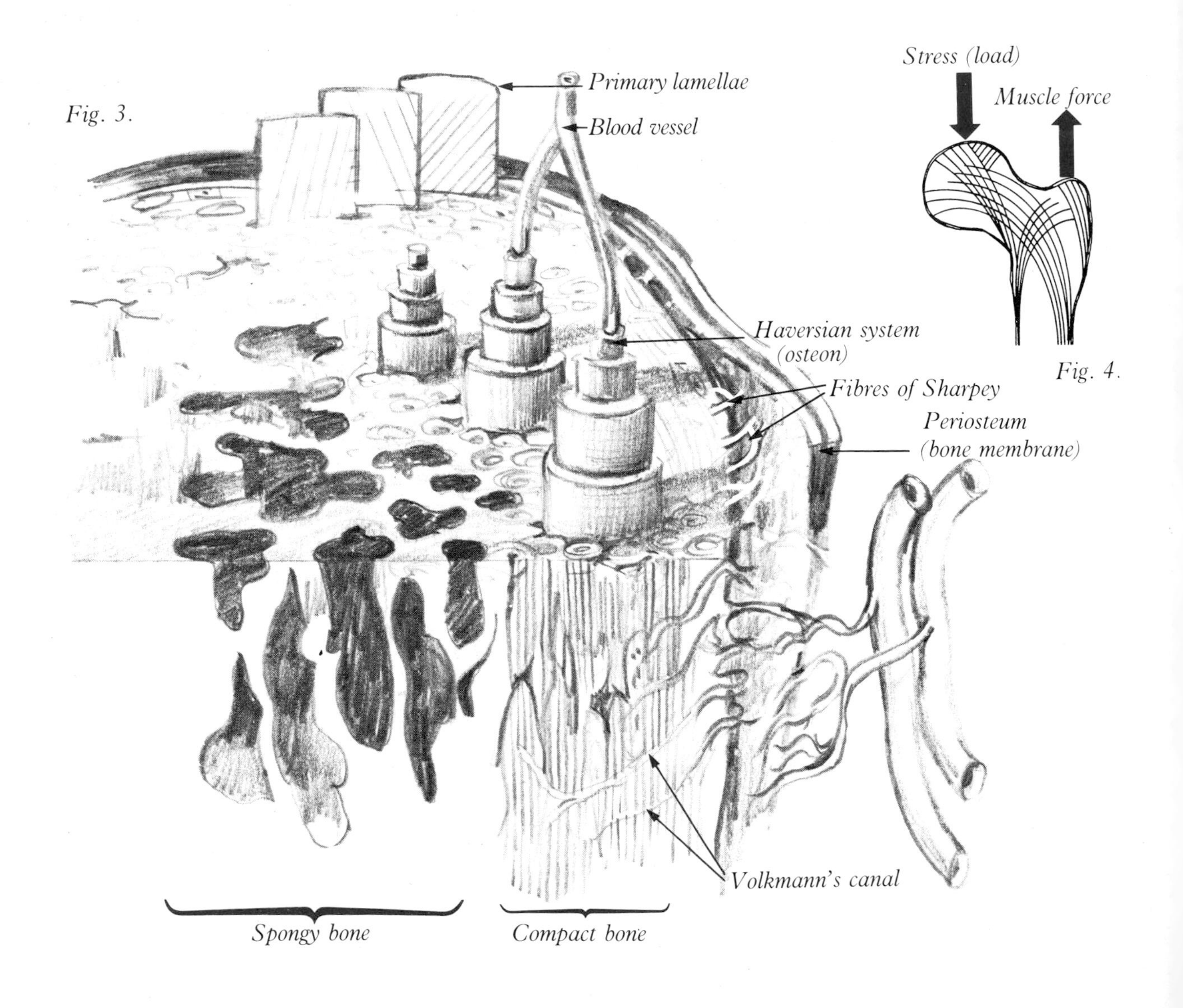

Fig. 3.

Fig. 4.

(giving the bone its hardness) and organic salts (giving the bone its elasticity). The ratio between the inorganic and organic salts is 1 : 1 at birth, but changes to 7 : 1 by the time the body is 60 - 70 years old. This explains the skeleton's elasticity in youth and its frailty in old age. Also shown in Fig. 3 is how the bone cells are arranged circularly in several layers around a so-called Haversian canal through which a small blood vessel runs. This blood vessel supplies the many layers with nourishment. Such a system is called a Haversian System, or osteon.

The outer layers of the bone form a system of longitudinal lamellae. The tough collagenous fibres run in different directions in each layer, which considerably enhances the strength of the bone. The bone within the lamellae is made up of osteons. Farthest in towards the medullary cavity, the compact bone becomes so-called spongy tissue. Between the compact and spongy tissues are reinforcing bars or braces which give the bone its great strength.

Figure 4 shows how these bars are constructed in the neck of the thigh bone (femoral neck) in order to withstand the considerable stress to which this part of the bone is subjected.

The muscle tendons and ligaments are attached to the bone by collagenous fibres growing through the bone membrane and into the compact bone tissue. When subjected to severe stress a tendon may hold, but its attachment to the bone may give way. A bit of the bone may thus be torn off or avulsed.

Where the muscle's tendon of attachment passes through the bone membrane — which is rich in nerve and blood vessels — an unfavourable stress can irritate the membrane and lead to periostitis (inflammation of the periosteum) (see p.61).

In the deeper layer of the bone membrane there are many bone-forming cells (osteoblasts) which are responsible for the repair of broken bones (fractures). The bone is supplied with nutrients by the vast number of blood vessels which penetrate the compact tissue via the bone membrane. The vessels then ramify and reach the different parts of the bone through the Haversian canals. The ability of the bone to repair itself is directly dependant upon the adequacy of the blood supply to the injured area. For example, when one of the bones of the hand (metacarpals) is injured it often takes a long time to heal as there are few blood vessels in the area and, moreover, those that do exist are usually injured in the fracture.

Scientific investigation has revealed that the number of capillaries supplying muscles and bone increases if the muscles and bones are regularly subjected to stress (training). This could explain the observation that injuries sustained by well-trained people heal much faster than those sustained by untrained people.

B. Joints

The different parts of the skeleton are connected either by attachments such as membranes or by joints.

Attachments (membranes, ligaments, discs)

Between the the shin bone (tibia) and the calf bone (fibula) there is a uniting membrane (interosseous membrane) formed of collagenous fibrous tissues. It has two functions: (1) it serves as an origin for many of the muscles of the lower leg (see p.60), and (2) it transmits stress from the shin bone to the calf bone. For example, when landing from a jump, force is exerted on the ankle bone (talus) and up through the tibia, but is then transmitted via the membrane to the fibula. (The tibia is relieved of stress.)

Fig. 5.

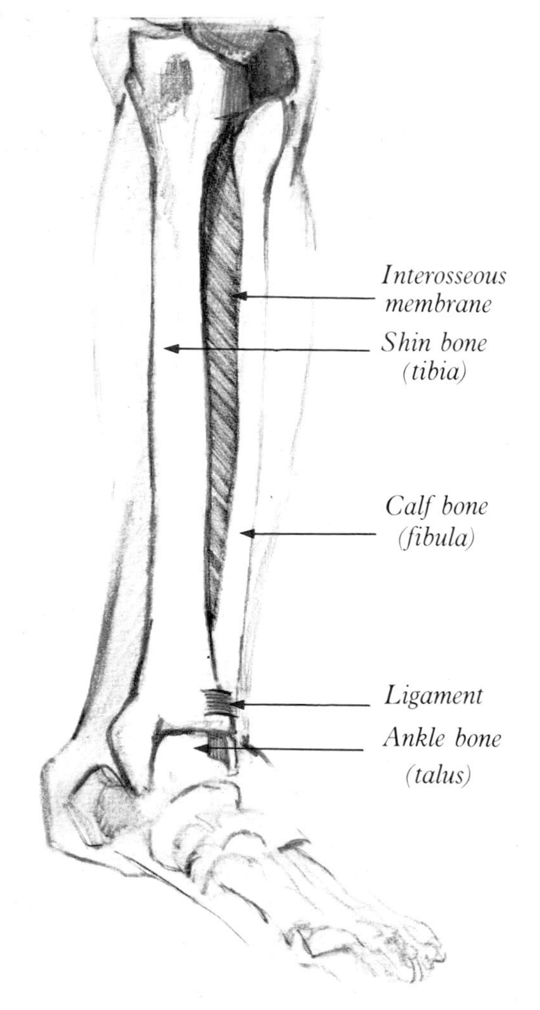

The tibia and fibula are also united by two strong bands of fibrous tissue at the ankle (inferior tibio-fibular joint — a syndesmosis). Such powerful, clearly distinguishable bands are called ligaments.

If the foot is violently pressed up against the lower leg (extension or dorsiflexion) the talus may wedge in between the fibula and tibia with such force that the anterior of the two ligaments may tear (rupture). The lower part of the membrane may also rupture.

Fig. 6.

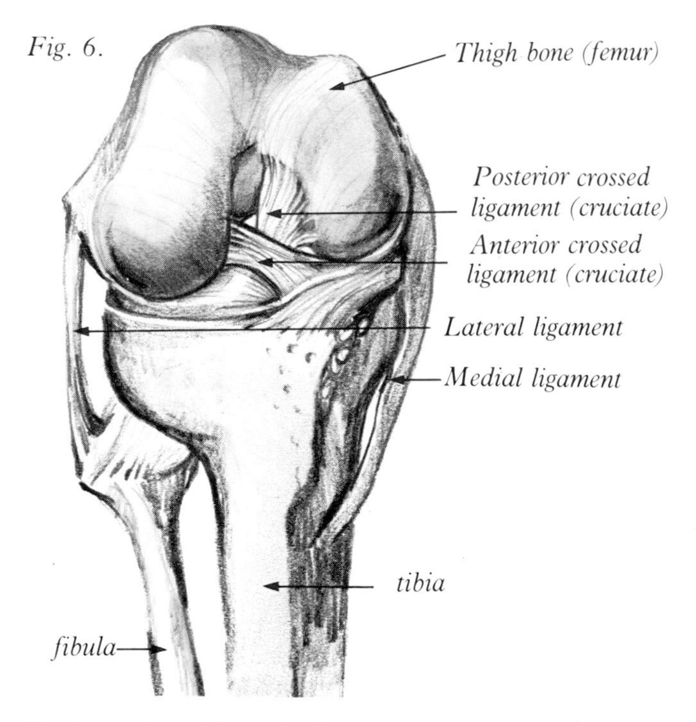

The knee viewed from the front (without a kneecap)

Other examples of clearly distinguishable free ligaments are the crossed, or cruciate, ligaments which cross each other inside the knee joint.

Another type of ligament provides reinforcement in the capsule which encloses a joint. This ligament is responsible for blocking those movements whose deviations are too great. It also restricts movements in certain directions.

The strongest of all the ligaments in the body (iliofemoral ligament) is a thickened band of fibres situated on the anterior aspect of the hip capsule. It restricts excessive backward swings of the legs. The strength of a ligament or tendon can be 5000 – 10000 N/cm^2. The above-mentioned ligament in an adult person can withstand a stress of 3000 N.

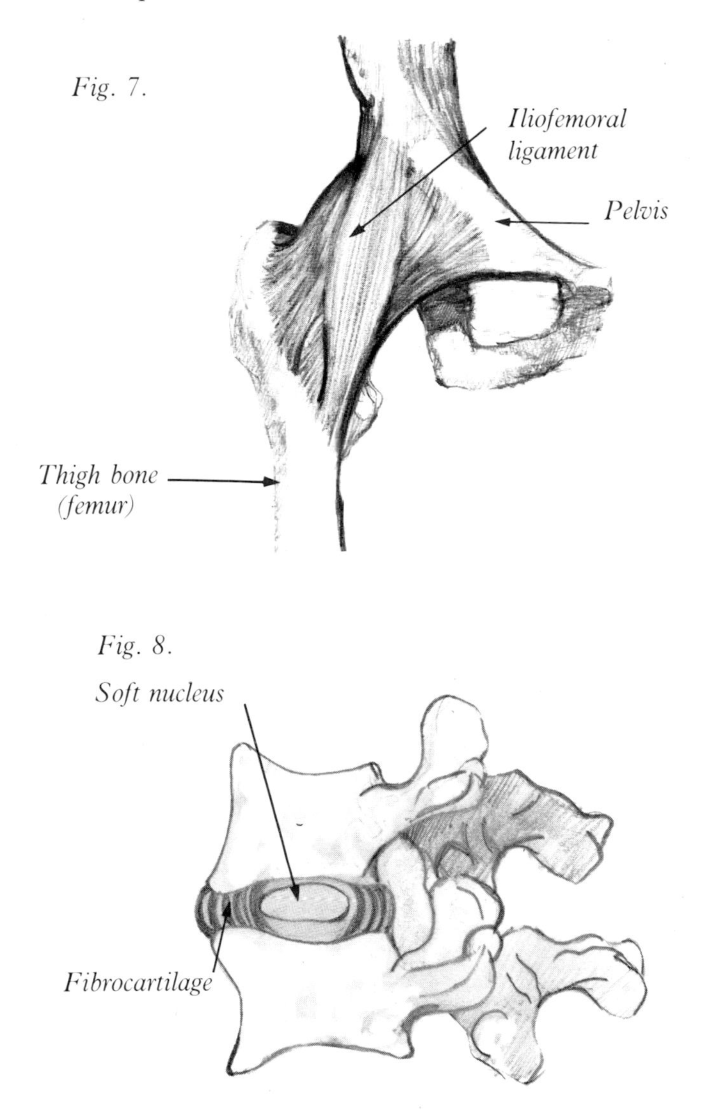

Fig. 7.

Fig. 8.

Fig. 9.

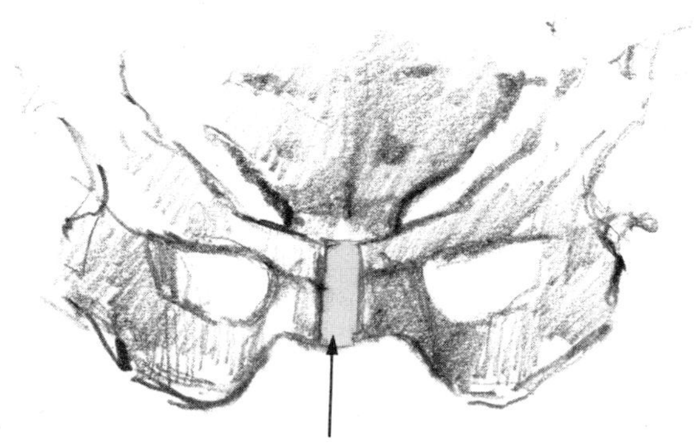

Symphysis pubis (fibrocartilage)

Another type of attachment can be found between the vertebrae of the spinal column. It is composed not only of collagenous fibres, but also of cartilage cells; and in its centre is a soft nucleus. It is called a secondary cartilaginous joint. A further example of this type of cartilaginous joint is the symphysis pubis connecting the pubic bones of the pelvis.

Additional examples of joint types are that of the epiphyseal cartilage (a primary cartilaginous joint) (see p.7), and the fibrous joint between the flat bones of the skull (sutures).

Fig. 10.

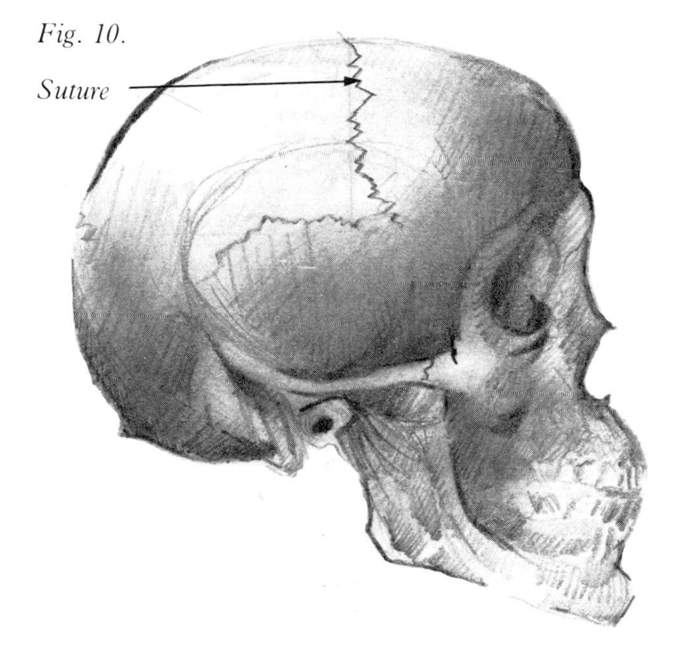

Joints (synovial joints)

The parts of the bones of a synovial joint are always enclosed in a joint capsule and they are always covered with articular cartilage. The outer layer of the capsule is formed by collagen fibres which are highly resistant to force. The powerful reinforcements of the capsule wall are called ligaments (see Figure 7). They receive their names according to their position or according to the bones they connect.

In the inner layer of the capsule there are cells that produce a fluid containing albumin. This fluid acts as a lubricant and provides the cartilage cells with nutrients. It is also called synovial fluid and the membrane which lines the capsule is called the synovial membrane. The inner and outer layers are separated by a thin layer of fat (Figure 11).

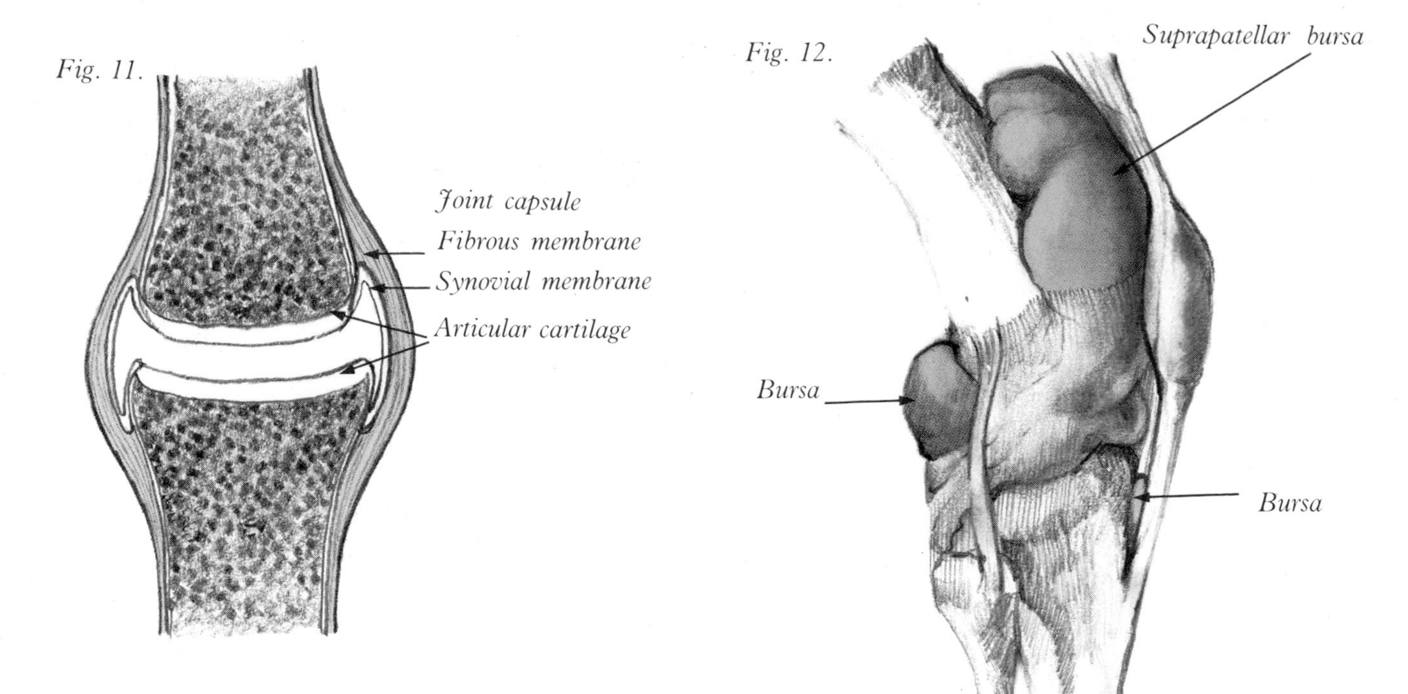

Fig. 11.

Fig. 12.

The thickness of the cartilage of a joint depends on the stress to which it is normally subjected. The cartilage is able to absorb certain substances from the synovial fluid and swell temporarily. If we measure the size of the articular cartilage after a period of warming up, we would find that it had thickened. The thickening is only temporary, lasting for 10–30 minutes after the activity has ceased. Prolonged training causes the cartilage to thicken by the formation of additional cartilage cells. Severe or uneven stress can wear the cartilage down, resulting in seriously restricted movement at the joint.

Little sacs of synovial fluid (bursae) are found associated with certain joints; they are formed in the same way as the joint capsule (Figure 12).

The bursa's inner layer (synovial membrane) causes it to form a displacement cushion with very low friction.

The bursa's task is, above all, to prevent wear of the different structures which glide against each other. However, it also produces synovial fluid in those cases when it is connected with a joint.

The largest bursa of the body (the suprapatellar bursa) is situated between the thigh bone (or femur) and the knee extensors (the quadriceps muscle). When the knee is subjected to severe stress, the bursa becomes sore. It reacts by producing extra quantities of synovial fluid. This leads to a swelling of the knee which prevents further stress (water on the knee).

Bursae can be found in many different places in the body. For example, between muscles, between tendons and muscles, and between tendons and bone, i.e. wherever wear and tear is likely.

The bones that make up a joint generally fit together well. Usually one of the bones is convex (the head) and the other is concave (the socket or depression). If, however, the bones do not fit together well, the irregularities are evened out by extra layers of fibrocartilage. These inclusions are called menisci if they only partly subdivide a joint cavity. If the joint is completely partitioned into two separate parts, the layer of fibrocartilage is called a disc.

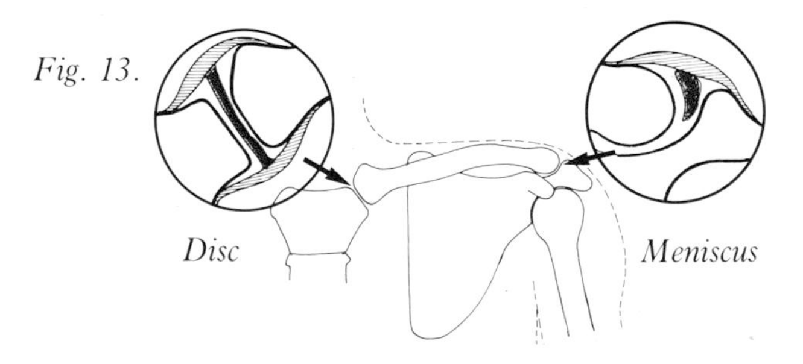

Fig. 13.

In sports injuries, reference to menisci usually refer to those of the knee, although one can also find menisci between other bones of the body. For example, between the collar bone and the shoulder blade (Figure 13).

Different types of joints

The function of joints is usually described with the aid of mechanical models. However, there is not always such a close resemblance between mechanical models and the actual joints of the human body. The diagrams below show different types of joints along with the parts of the body where they can be found. In addition, illustrations are given of the type of movement allowed by these joints.

Fig. 14.

Hinge joint (movement in one plane only)

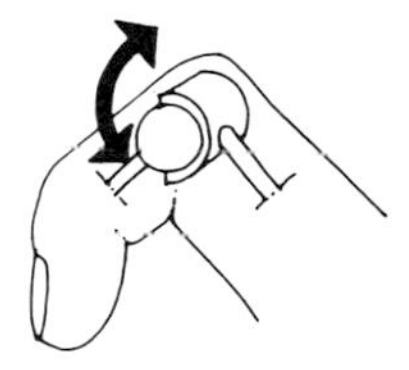

Stretching (extension) *Bending (flexion)*

Fig. 15.

Pivot joint (movement round one axis only)

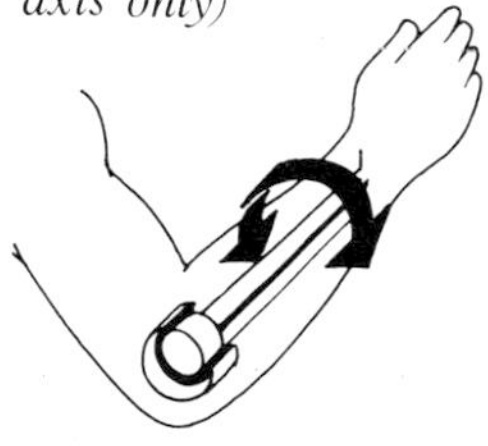

Inward rotation (pronation) *Outward rotation (supination)*

Fig. 16.

Saddle joint (movements round two axes)

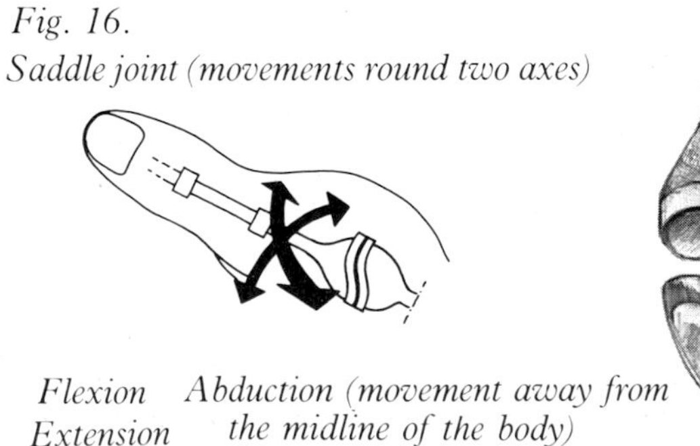

Flexion
Extension

Abduction (movement away from the midline of the body)
Adduction (movement towards the midline of the body)

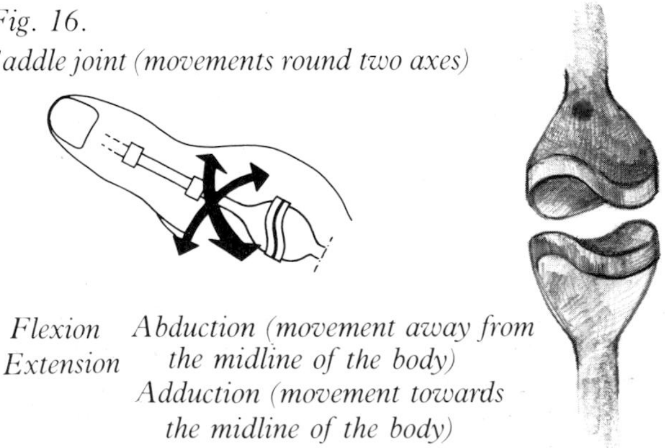

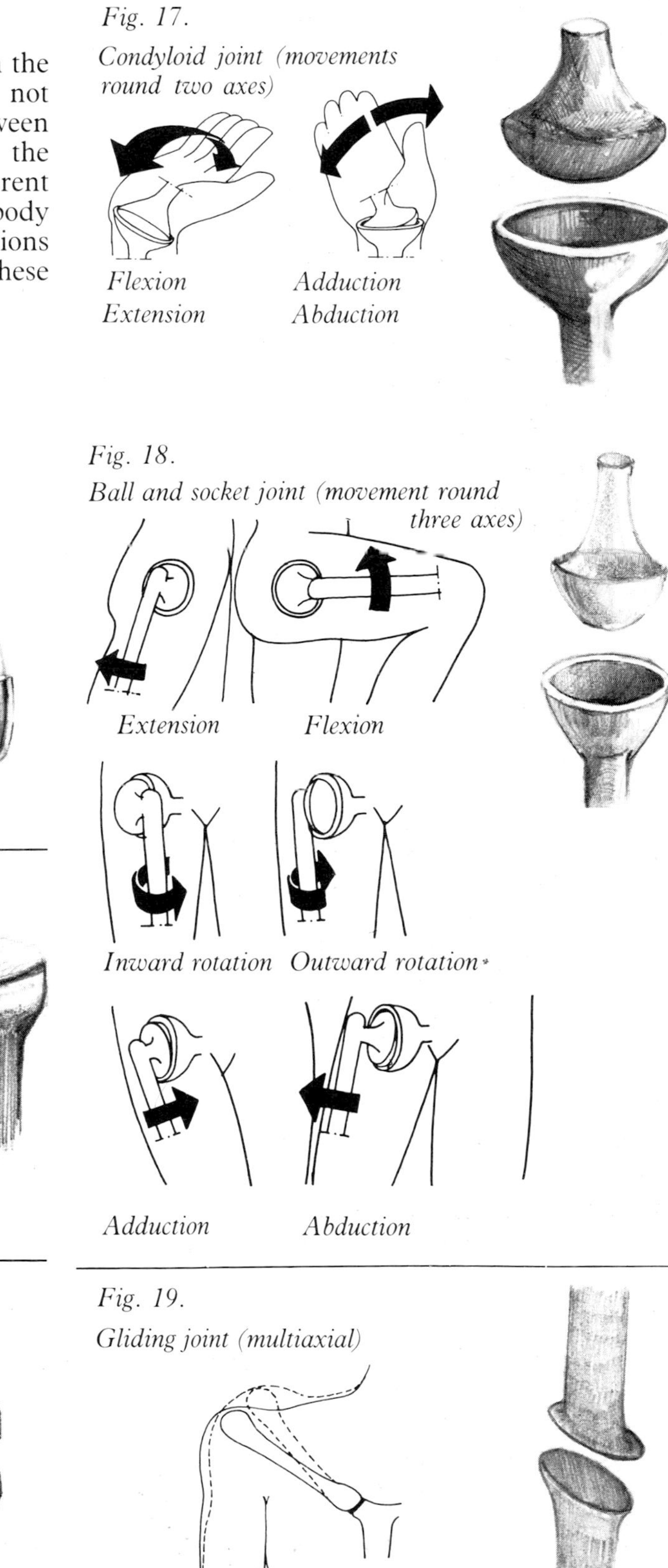

Fig. 17.

Condyloid joint (movements round two axes)

Flexion
Extension

Adduction
Abduction

Fig. 18.

Ball and socket joint (movement round three axes)

Extension *Flexion*

Inward rotation *Outward rotation*

Adduction *Abduction*

Fig. 19.

Gliding joint (multiaxial)

Small versatile movements

Because the capsular ligament around a gliding joint is almost always taut, the movements allowed at such joints are described as small but multidirectional (hence multiaxial).

Many of the body's joints cannot be compared with any of the above models. Instead, a combination of two models or one of the models with some modification is used for descriptive purposes. For example, the knee joint is a combination of a hinge and pivot joint. At that joint, flexion and extension are possible as well as inward and outward rotation of the lower leg, but the leg can only be rotated when the knee is bent.

The knuckle joints are modified ball and socket joints. The head of one bone and the socket of the other correspond almost exactly to the ball and socket model, but the capsular ligament prevents certain movements and there are no muscles for rotating the finger.

Fig. 20.

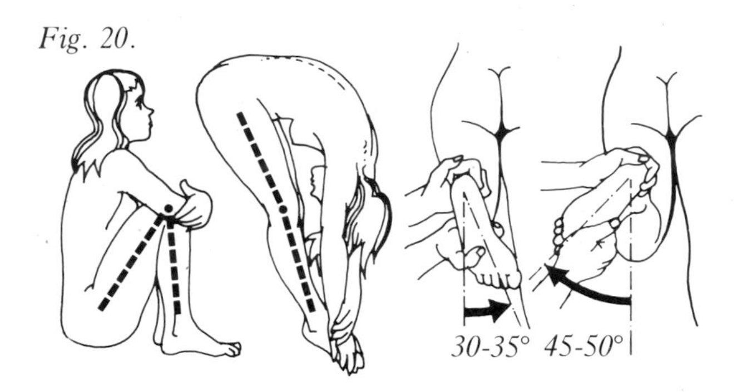

Of all the muscles that protect the joint from injury, the muscles directly surrounding it are the most important. A strong and supple musculature is always the best protection against joint injuries. Violent physical activity can lead to different kinds of muscle injuries. Muscles can be overstretched, partly ruptured (partial rupture) or even completely ruptured. When a ligament is stretched, it usually returns to its original length and function after some weeks of rest.

If the same ligament is repeatedly stretched, the result can be a slack non-functional ligament. The joint then only fits loosely together, which can lead to serious injury. Such a ligament should be shortened by surgery. Ligaments that are completely torn should be sewn together. Partially torn ligaments can be operated on, i.e. sewn together, but if the torn parts are in contact they can heal if the affected area is kept still. In such cases the joint should be set. In recent years it has been shown that even a completely torn heel or Achilles tendon heals if the ankle is set in such a position that the parts of the tendon lie against each other.

A ligament can be made slack and unstressed by taping. The aim of taping a joint is to prevent movements that can strain a weak or injured ligament. Taping in athletics is defensible after injury has been sustained or as part of an effort to prevent injury when it is known that the risk of injury is high (e.g. in vigorous football training, or other competitive situations). However, an athlete should avoid the habitual use of taping during training as this could lead to an injury of the skin. Moreover, the unloaded ligament becomes accustomed to external help after a while, and is thus weakened.

C. Muscles

There are three different types of muscles in the body: smooth muscle, cardiac muscle and skeletal muscle. This book will only deal with the last of these, which is also called striated or striped muscle.

A skeletal muscle is surrounded by a layer of connective tissue. The connective tissue is built up in the same way as the outer layer of a joint capsule (see p.12). Its task is to provide a surface against which the surrounding muscles can glide, and it gives a muscle its form.

The layer of connective tissue is also called the muscle's fascia or epimysium. The connective tissue is formed mainly of collagen fibres. Looking at a section of a muscle with the naked eye, we see that it is made up of small cell bundles (fasciculi). Each of these is surrounded by a thin layer of connective tissue whose latin name is perimysium. In this layer of connective tissue — which is made up of both collagenous and elastic fibres — the nerve and blood vessels branch off before finally reaching the actual muscle cells. Under the microscope we can see that each fasciculus consists of a number of muscle cells. Each cell is surrounded by a very thin layer of connective tissue which is called endomysium (endo = final, mysium = muscle).

The muscle cell is also called a muscle fibre, thus a muscle fibre consists of a single cell. The structure and function of muscle cells is described very thoroughly in the majority of books on physiology. The following account is therefore brief.

When examined under the microscope, the muscle cell is seen to be composed of small components called muscle fibrils or myofibrils (fibril = little fibre). The fibrils lie in parallel and give the muscle cell a striated appearance. This is because the fibrils are made up of smaller components, myofilaments (filaments = smaller than fibrils), which are regularly aligned.

Myofilaments are chains of protein molecules. The striated appearance is due to the presence of two types of myofilament, namely, actin (which is thinner and therefore more transparent) and myosin (which is thicker and is responsible for the darker bands).

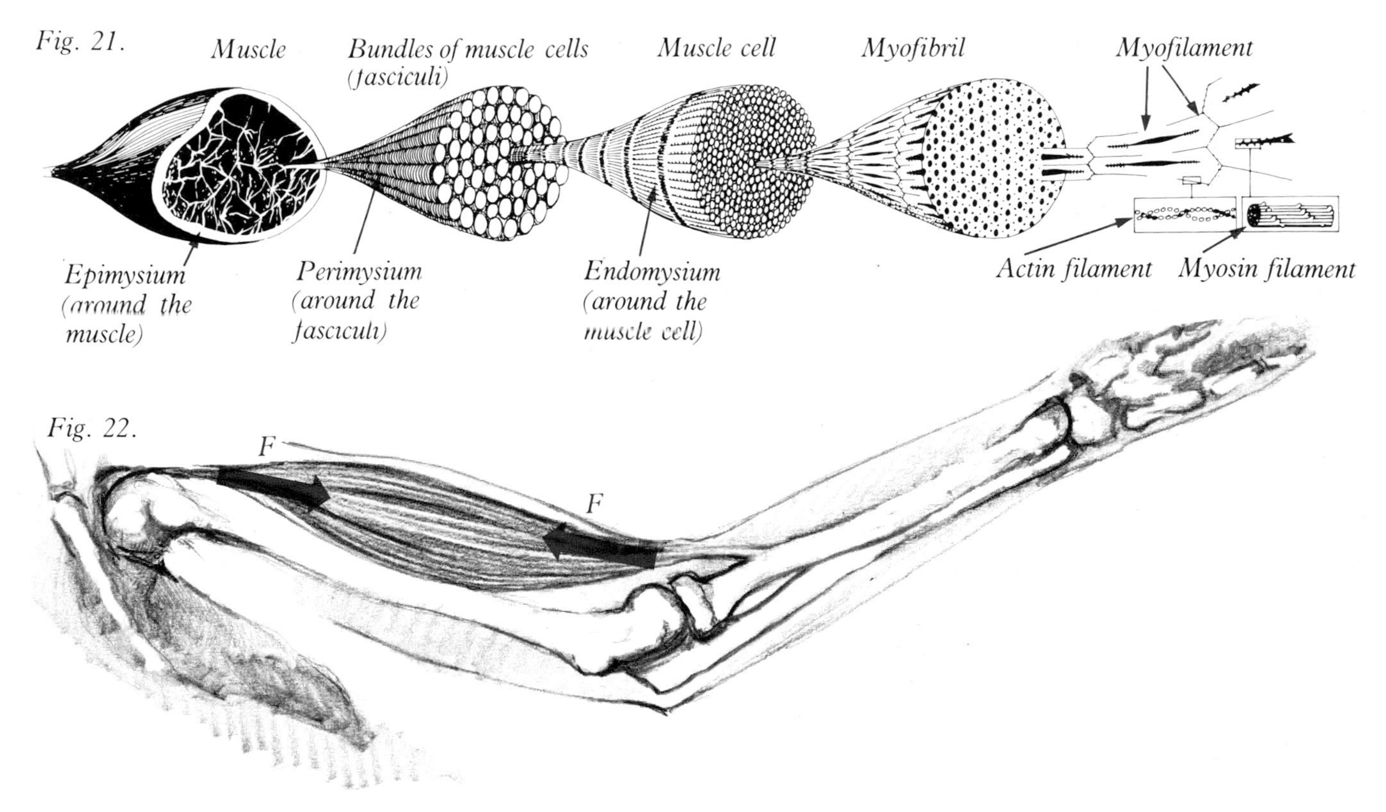

Fig. 21.

Fig. 22.

When the muscle contracts, the actin filaments move between the myosin filaments. As a consequence, the myofibrils shorten and thicken.

The connective tissue surrounding the muscle extends and is continuous with the muscle's tendon. When a muscle contracts, it produces a force *(F)* which affects the origin and insertion of a muscle equally, but in opposite directions.

The muscles of the body have very different shapes. Figure 23 shows the most common variations.
A muscle can develop a maximal force of about 50 N/cm^2 of the muscle's cross section. The cross section referred to here is physiological, and it is defined in the following way: if the cells of a muscle run longitudinally, the geometrical cross section of the muscle *(A)* will provide a measurement of how many actin and myosin filaments are contained in the muscle. If the area is 6 cm^2, the maximal force of contraction will be 6 x 50 = 300 N. If the muscle cells run diagonally in relation to the muscle's longitudinal direction, then the areas A_1 and A_2 must be measured in order to arrive at the total number of actin and myosin filaments contained in the muscle.

Fig. 23.

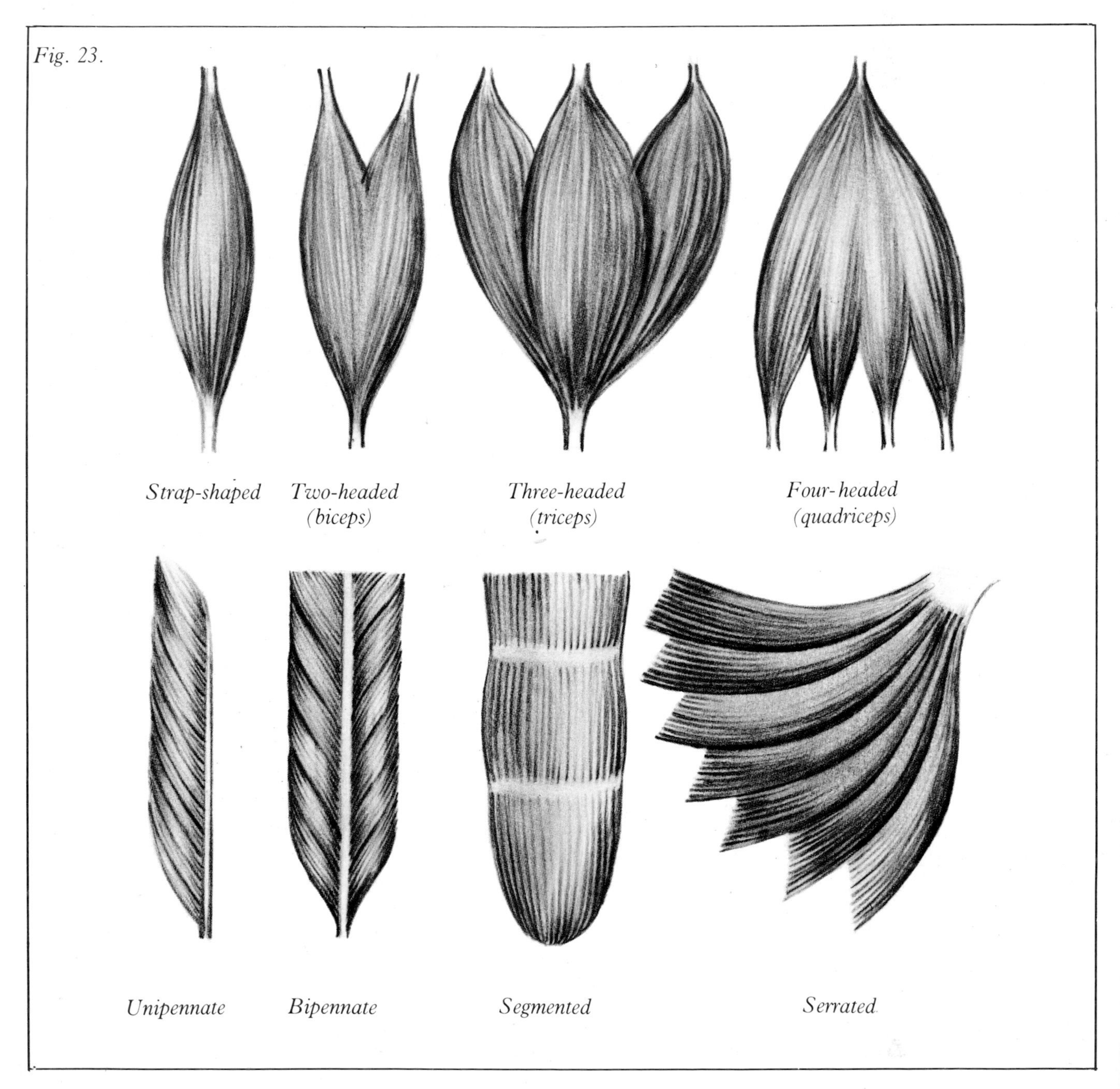

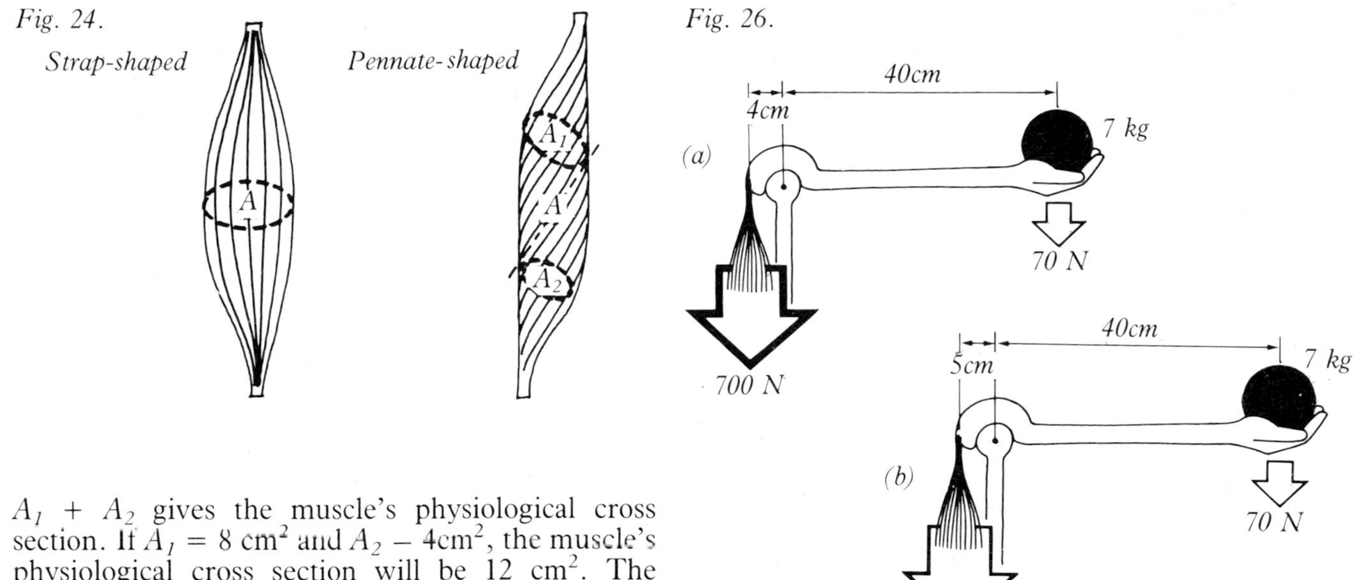

$A_1 + A_2$ gives the muscle's physiological cross section. If $A_1 = 8\ cm^2$ and $A_2 = 4cm^2$, the muscle's physiological cross section will be 12 cm^2. The muscle's maximal strength is 12 x 50 = 600 N. This muscle is thus considerably stronger than the strap-shaped muscle, even if both muscles have equal mass. A muscle cell can shorten its length by about 50%. The strap-shaped muscle can therefore be shortened over a greater distance than the pennate muscles.

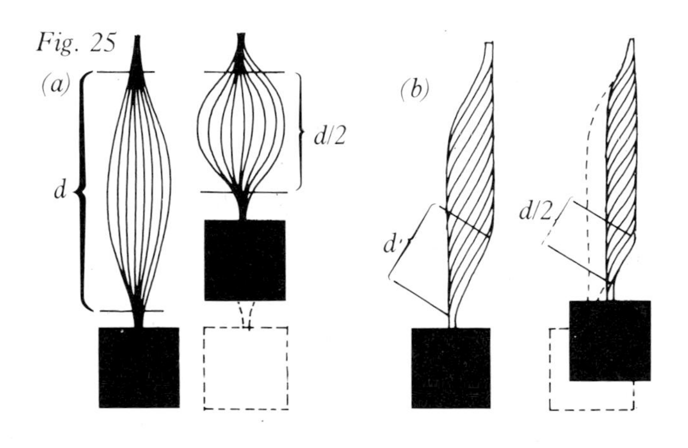

The strap-shaped muscle is found in places where it is necessary to execute large ranges of movement quickly. Pennate muscles, on the other hand, can be found where movements over a small range but of great strength are required.

The force of a muscle is dependent on its physiological cross section. Its ability to contract depends on the length of its fleshy part measured in the direction of the cells. In order to assess the effect of a muscle we must also know where it is attached in relation to the joint.

The diagrams above show that if a muscle is attached 4 cm from the joint axis, its force must be 700 N if it is to support a weight of 7 kg located 40 cm from the joint (70 x 4 = 7 x 40). If, instead, the same muscle were attached 5 cm from the joint, then its force need only be 560 N (56 x 5 = 7 x 40). The ability of a muscle to lift a heavy object is thus dependent on two factors: its physiological cross section and its position in relation to the joint. A muscle's strength is more adequately described when we know its ability to develop the moment of force *(M)* as well as its force of contraction. (The concept "moment of force" is described on p.30.)

In order to analyse a movement, it should be known that a muscle's force of contraction depends on its contracted length as compared to its length at rest. Suppose we want to investigate the ability of a muscle bundle to contract. The muscle bundle's resting length is d_0 cm. The muscle bundle is attached to two plates which are connected to a gauge. The gauge shows the force with which the plates are drawn towards each other.

If the distance between the plates is d_0, then the force is 0 (F_0). (The beginning of the broken-lined curve in Figure 32). If the muscle bundle is drawn out, the indicator will respond. This is because the elastic components of the muscle manifest their natural tendency to remain at rest d_0. The broken-lined curve below shows the magnitude of the force with which the muscle would return to its original resting length when it is stretched. (The muscle is passive, i.e. it does not contract.)

Fig. 27. F_0

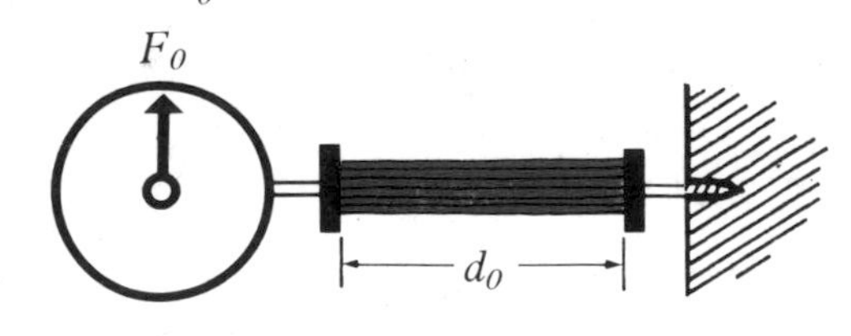

Fig. 28. F_1

If the distance between the plates is extended to d_1, then the force will be F_1.

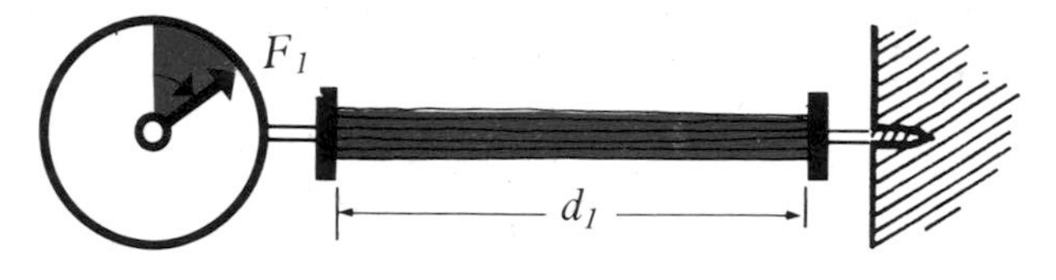

In order to examine which forces are developed when a muscle contracts, we do the following: via an electrical impulse, we incite the muscle cells to contract, and the force is gauged.

If the electrical impulse is given when the distance between the plates is d_0, then the force will measure F_2 (the absolute value will depend on the thickness of the muscle bundle).

If the distance between the plates is less than d_0, then the muscle force *(F_3)* which could be exerted on the plates, would be less than F_2.

Fig. 29. F_2

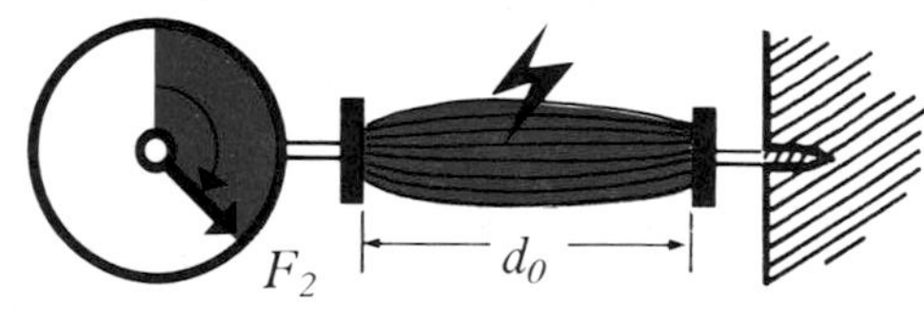

If the distance is reduced to about 50% of the muscle bundle's length *(d_0)*, then during contraction the origin and insertion will only just reach the plates, i.e. the force will be 0.

Fig. 30. F_3

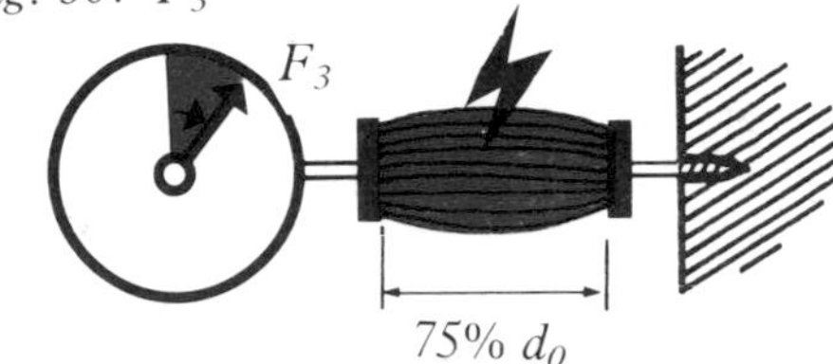

If the muscle is drawn out slightly beyond its resting length and thereafter stimulated electrically, it will be seen that its ability to contract will be greater than it is at d_0, i.e. F_2. This greater value is due to the combination of elastic forces and the forces brought into action by the electrical impulse. The maximum value *(F_4)* is reached when the muscle is extended to about 120% of its resting length.

Fig. 31. F_4

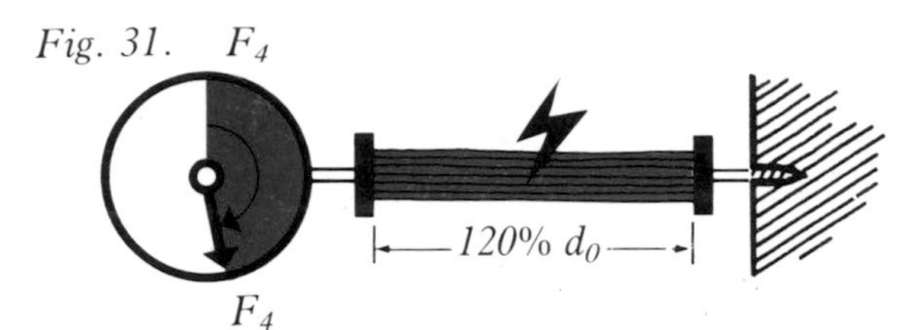

Fig. 32.

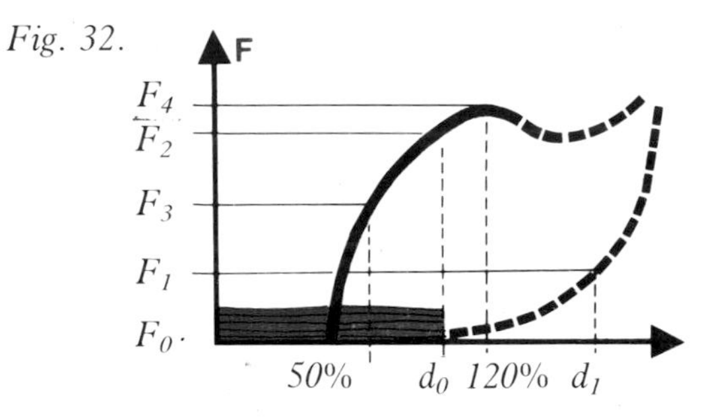

If the muscle is drawn out more than 120%, then its force of contraction decreases again. This is because the actin and myosin filaments are pulled so far apart that the muscle's capacity for voluntary contraction decreases more than the elastic force of contraction increases.

When we want to obtain the maximal strength of a movement, a "good technique" for doing this involves working the muscles under the most favourable conditions possible.

The hip muscle that is engaged when the leg is pushed forcefully backwards is the large buttock muscle (gluteus maximus).

When a person walks on level ground he does not need to push off from the ground with excess force, but if he encounters a steep ascent, he automatically bends forward. Thus, the muscle is lengthened so that the force developed is greater (120% as shown above).

Fig. 33.

Think about what a person does to increase his speed when skating, running or cycling. In order to give a football a mighty kick, for example, the hip must start from a position well back as an important knee extensor muscle (p.49) passes over the hip joint. If the muscle is required to develop much force, then the distance between its origin and insertion must be greater than d_0.

Fig. 34.

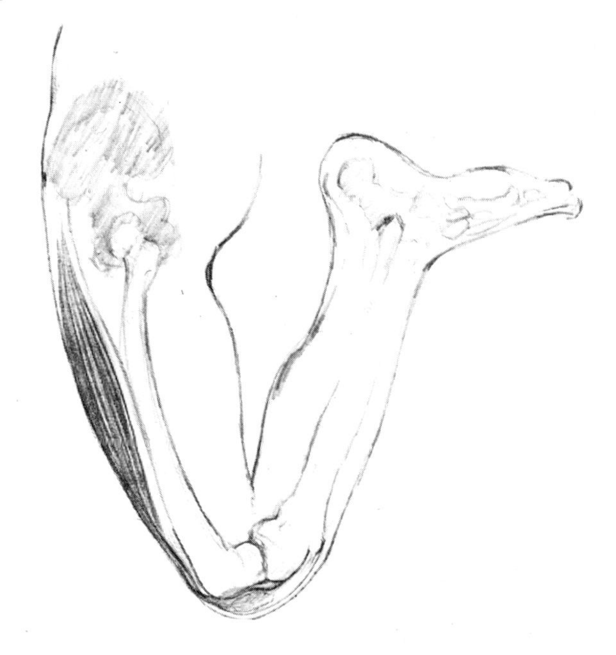

For efficient throwing it is necessary to lengthen an important group of muscles, namely, the pectoralis major (p.82). It is stretched in relation to the arm, when the trunk rotates fully in the opposite direction. At the same time, the chest is expanded by inhaling deeply.

Fig. 35.

Different types of muscular work

When observing the way muscles work, it is usual to distinguish between dynamic and static work.

Dynamic work means that the origin and insertion of a muscle are forcefully affected by changes in muscle length. If the muscle force causes the origin and insertion to move towards each other, we say that the muscle works concentrically (the muscle is shortened, contracted). If the muscle force is exerted while the origin and insertion are receding from each other (i.e. the muscle tries to halt a movement in a joint), the muscle is said to work eccentrically (although the muscle tries to shorten, it is actually lengthened by external forces).

When a muscle contracts without any movement taking place in the joint, it is said that the muscle works statically (or isometrically).

Fig. 36.

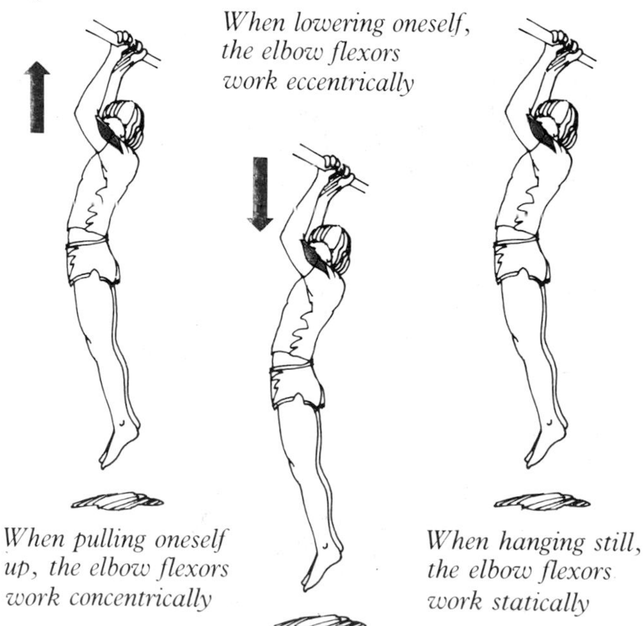

Fig. 37.

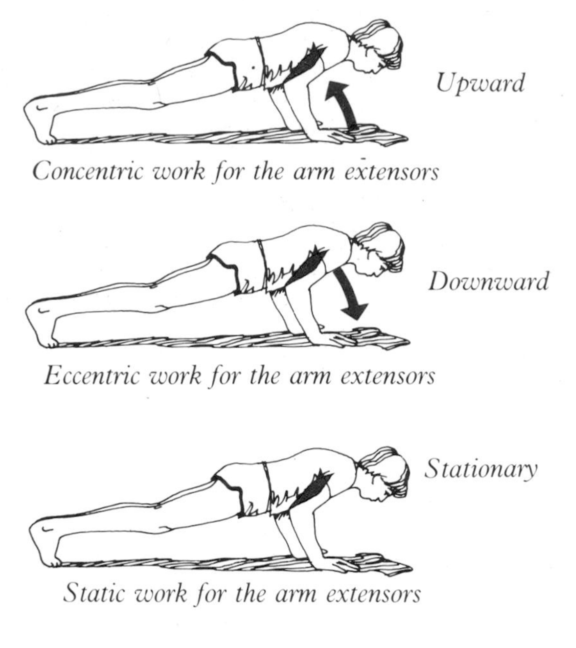

Fig. 39.

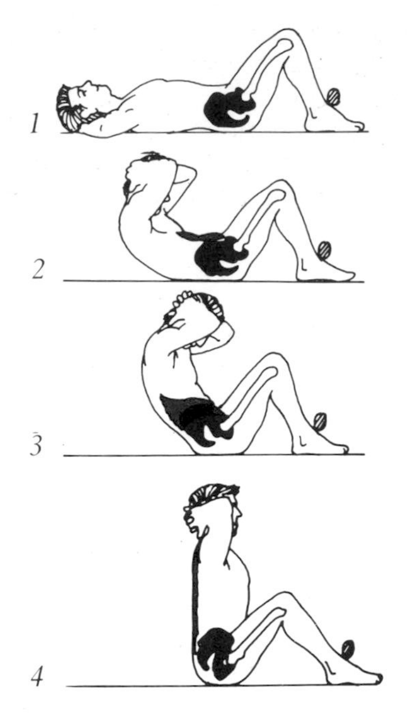

Figure 38 shows a training exercise that provides concentric work for the lumbar muscles (the movement takes place mainly at the small of the back), and static work for the neck and chest muscles. During descent, the lumbar muscles work eccentrically in an effort to halt the movement. (For the back muscles, see p.68.)

Fig. 38.

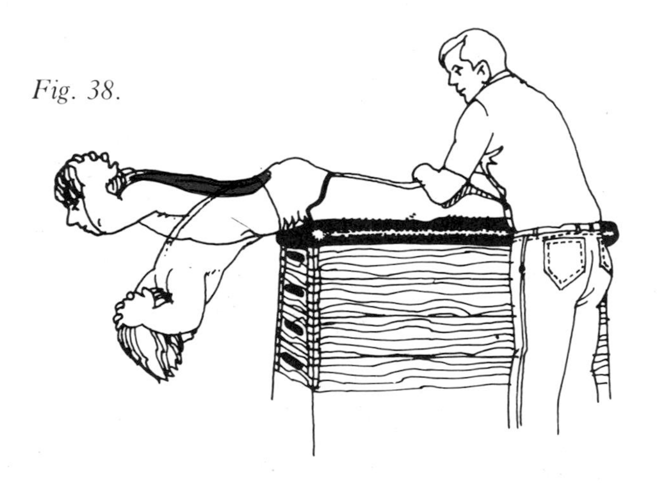

Figure 39 shows what happens when a person sits up from a lying position.
1–2. Concentric work primarily for the straight abdominal muscle (p.71).
2–3. Static work for the abdominal muscles, concentric for the hip flexors (p.43).
3–4. Almost no work for the abdominal muscles. Concentric work for the hip flexors.
4. Static work for the back extensors.

It is easier to analyse the exercises that are used in human motion, competitive sports, strength training, flexibility training, etc., when we know the origin and insertion of muscles. Physiological studies have shown that if a muscle is trained concentrically, its ability to work statically and eccentrically is not appreciably increased. We can assume that the reason for this lies in the fact that the muscle's ability to work is dependent partly on its mass, and partly on its supply of nerve impulses. Both these functions must operate for the muscle to be effective. A good athlete or trainer ought to be capable of analysing his particular branch of athletics and designing exercises to meet the demands of the movements involved.

Fig. 40.

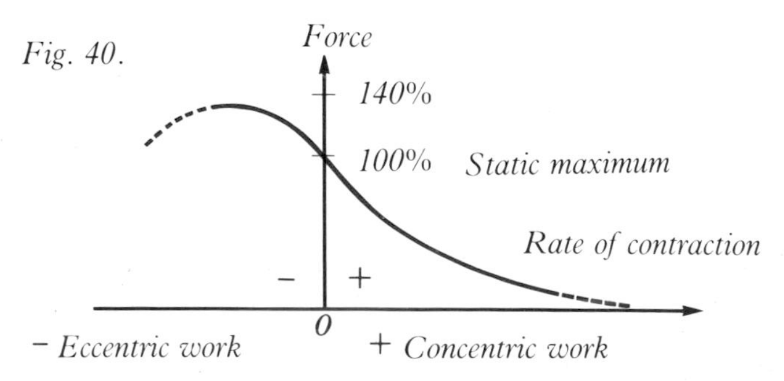

+ Means that the point of origin pulls towards the point of insertion (concentric work).
0 Means unchanged distance between origin and insertion (static work).
− Means that the point of origin is pulled away from the point of insertion (eccentric work).

When we investigate a muscle's ability to develop force, we see that the muscle becomes up to 40% stronger when made to work at its eccentric maximum compared with its static maximum. Its concentric ability to develop force decreases as the velocity with which it must work increases.

Thus, to build up high tension in a muscle, we would force the muscle to halt a movement that actually requires the muscle to stretch (also called negative work). The following experiment demonstrates this function.

1. Stand still with your arms hanging vertically at your sides. Bend your knees at a 90° angle. Starting from this position (without swinging your arms or rocking), jump straight up and estimate the height you reach.

2. Stand with your arms hanging at your sides and your knees straight. First make a get-ready movement which will take you to the starting position of exercise 1, and then jump straight up. In this way you will jump higher. This is because, in the position where you changed direction (which is exactly the same as the starting position of exercise 1), the muscles used in the jump were already "charged" by having been forced to work eccentrically when halting at the 90° angle.

Fig. 41.

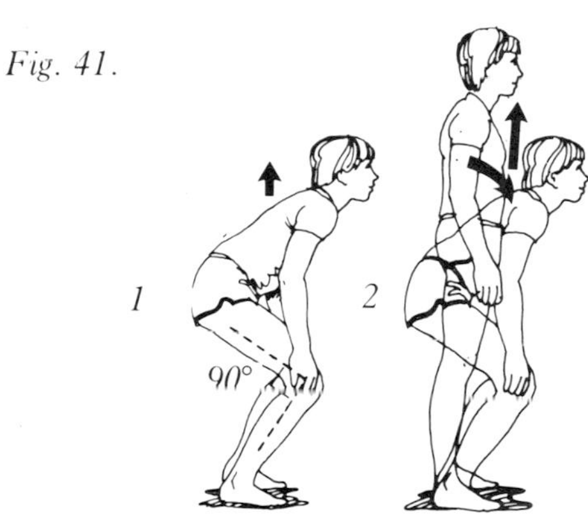

Thus, both jumps start from the same position but involve different muscle tension. An effective movement has a certain rhythm, i.e. pattern of movement, which eccentrically charges the different muscle groups taking part in the movement. It is said that a tennis serve, or a volleyball smash, requires "timing". Here, timing means that the entire pattern of movement is such that each group of muscles being used by the player works optimally.

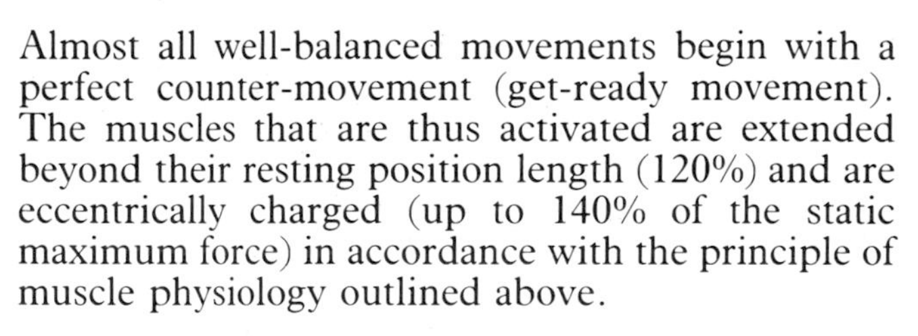

Almost all well-balanced movements begin with a perfect counter-movement (get-ready movement). The muscles that are thus activated are extended beyond their resting position length (120%) and are eccentrically charged (up to 140% of the static maximum force) in accordance with the principle of muscle physiology outlined above.

Knowledge of a muscle's origin, insertion and function, as well as external forces which can affect it (force of gravity, friction), allow us to analyse and understand the exercises used in athletics.

Muscles — Nervous System

The muscle's protective reflexes

There are two types of nerve cell that protect a muscle against unnecessary injuries: muscle spindles and tendon spindles. The muscle spindles are connected in parallel to different muscle cells throughout the muscle. The muscle spindles passively follow the movements of their adjacent muscle cells.

Fig. 42.

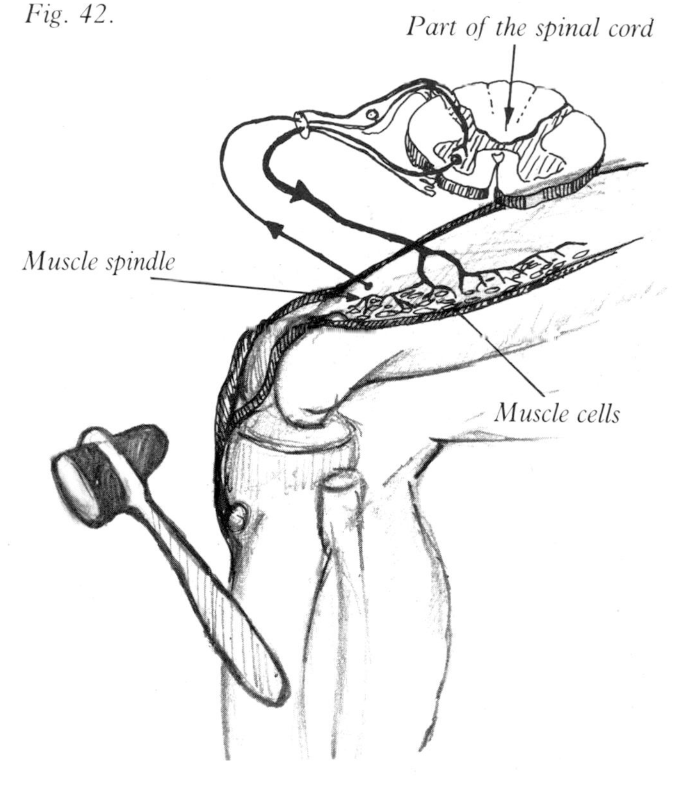

When the muscle cells stretch, so do the muscle spindles. If the muscle stretches so much as to run the risk of being ruptured, the muscle spindle responds by sending a signal to the muscle to contract. This keeps the muscle from being injured. This protective mechanism is called the stretch reflex.

When a doctor taps the ligament directly below the kneecap (ligamentum patellae) with his rubber mallet, the muscle cells in the knee extensor muscles stretch (see p.49). The muscle's reaction to this unexpected stretching is to protect itself by contracting — i.e. the knee jerks a little. The period of delay between the tap with the mallet and the kick is an indication of the time it takes the nerve impulse to travel from the muscle spindle to the spinal cord of the central nervous system (CNS) and back again to the muscle cells.

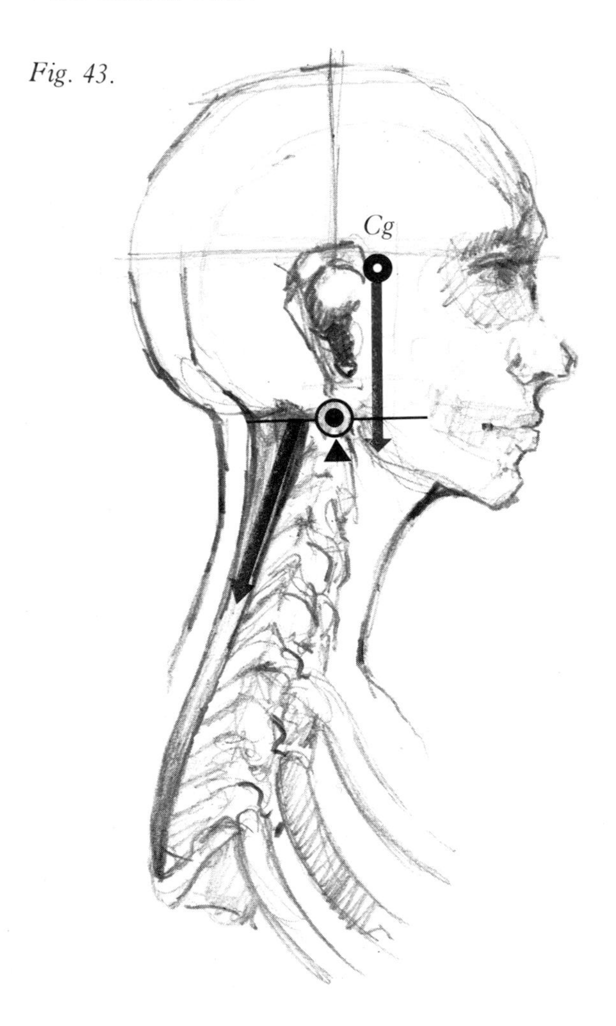

Fig. 43.

The centre of gravity of an adult's head lies above and in front of the joint which is anterior to the upper cervical vertebra (i.e. the uppermost vertebra of the neck). The head is held upright — despite its natural tendency to tilt forward — by the tension maintained by the muscles of the neck (see p.24). When a person falls asleep in a sitting position, the muscles of his neck relax, and his head falls forward. The muscle spindles stretch unexpectedly, which causes him to jerk his head up. This protective mechanism has probably saved the lives of many a tired motorist and kept many a bored listener awake.

The protective mechanism of the muscle spindle responds when the muscle spindle is stretched unexpectedly, but it does permit voluntary stretches which are not too sudden. The head may be allowed to fall forward–downward without the reflex being evoked. It has been shown that if a muscle first contracts and then stretches slowly, it can be made to extend a little further. These principles should be carefully adhered to when training flexibility and movement. They are described in more detail on page 24.

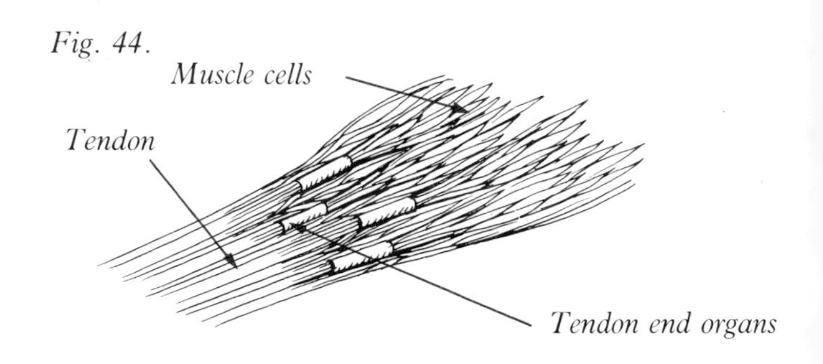

Fig. 44.

Tendon organs keep the CNS informed of the muscles' tension. If the tension is too high (and there is a risk of rupture) an impulse is sent from the tendon organs to the CNS and back to the muscle. The impulse inhibits the muscle from contraction which would lead to increased tension. This impulse is called the inhibitory impulse. Thus, the muscle relaxes and the tension decreases.

If a long-jumper stakes too much on his approach, the tension in his calf muscle can be too high. This may be caused partly by external forces, and partly by the contraction of the calf muscle. The muscle is inhibited and the jumper runs straight ahead — despite the fact that he is consciously charged for the jump.

Muscle spindles and tendon organs are a part of the body's most significant defence mechanisms. If they cease to function satisfactorily, the risk of injury increases considerably. This is particularly the case when an athlete is not warmed-up or is fatigued, as the protective reflex signals travel at a much slower rate then than when he is rested.

Motor unit

In order to judge the value of different strengthening exercises, one should be aware of the following.

A muscle is made up of a large number of muscle cells. These are an integral part of so-called motor units. Thus, the motor unit is composed of a nerve cell which has connections to the brain and which branches off into fine filaments terminating in the muscle. Each of these branches terminates in a single muscle cell. The number of cells in a motor unit depends on the degree of precision required from them. The muscles that are responsible for the movement of the eye, for example, contain five to ten cells in each motor unit. By comparison, the large muscle that forms the fleshy part of the buttock (p.39) is estimated to contain several thousand cells in each unit.

Fig. 45.

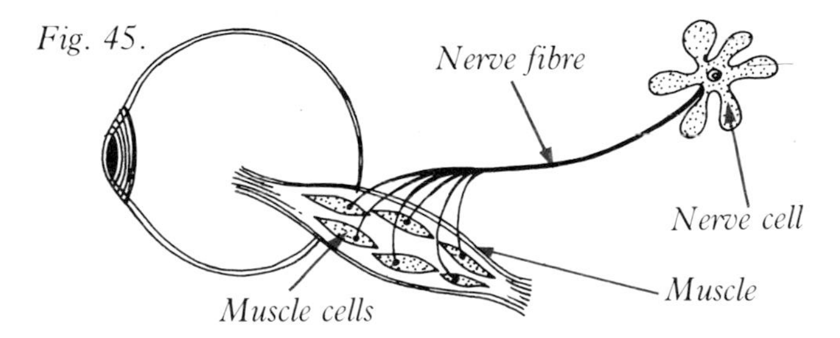

When the motor unit is put to work, all the cells that comprise it contract with maximal force. If a muscle is compelled to contract with a certain force, the work is carried out by a certain number of motor units. If the force of contraction is to increase, more motor units have to be engaged. It is always the same motor units that are used for coping with a light load, and it is always the same units that, thereafter, are engaged when the load increases. It is always the same motor units that are ultimately engaged when the work load becomes maximal.

Light work loads train only the muscle cells in the motor units that are first engaged. If we want to train the entire muscle, we must subject it to maximal stress.

There are two different types of muscle cell. They are called slow (type I) and fast (type II) muscle cells. The slow cells are characterised by the fact that they are supplied with energy via oxygen in the blood. The fast cells use mainly the energy stored in the muscle (as glucose) which can be transformed into mechanical energy without oxygen. A by-product of this process is lactic acid. Different people have different percentage distributions of these cell types. The usual proportions are 50% type I fibres and 50% type II fibres. There are, however, great individual differences. Moreover, different muscles have different compositions of fibre types.

In recent years it has been revealed that it would be appropriate to divide the type II cells into two sub-groups: type IIa and type IIb. Special training can change the character of type IIa cells in such a way that they become more like type I cells, i.e. they are supplied with energy via oxygen and are thus capable of greater endurance.

It has been shown that when a muscle is required to exert force the cell types are engaged in the order, type I cells, type IIa cells and, finally, type IIb cells. When little force is required, only the type I cells are trained.

Fig. 46.
Type I cells are characterised by endurance and little force.

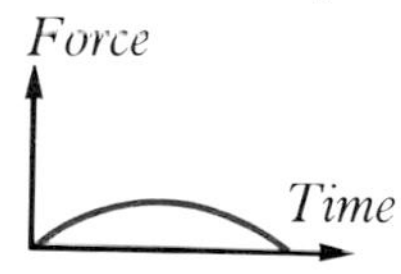

Type IIb cells are characterised by the great force they exert for short periods.

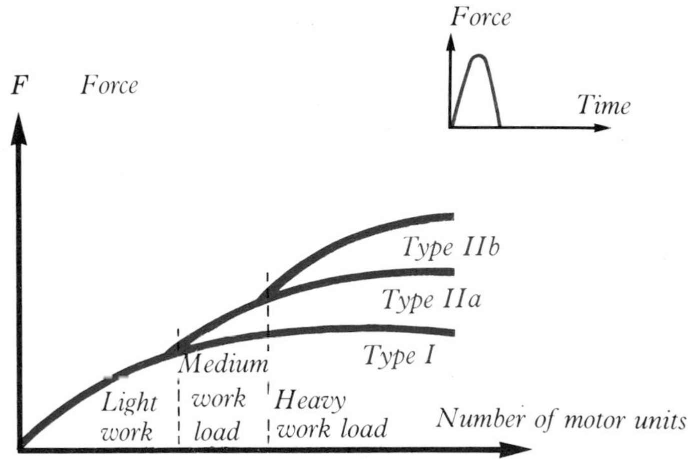

Type IIa cells have the best qualities of both the other types, i.e. great strength and good endurance. Knowledge of the above is important when we wish to train for a special type of sporting event. However, it should be made clear that the designation "slow" does not mean that the muscle cells contract so slowly that they do not contribute to the movements of sports. Such "fast" movements as a golf swing and tennis smash are actually very slow compared with the speed with which the type I cells contract.

D. Flexibility Training

Different types of stretching

In order to maintain the natural flexibility of the joints and reduce the risk of injury in athletics, the training programme should always contain suitable, and correctly executed flexibility exercises. A muscle that is only trained for strength becomes shorter. This, in turn, means that its range of movement is restricted, which in practice decreases its ability to utilize its increased force resources correctly. Exercises that are designed to train the strength of a group of muscles, should always be followed by stretching exercises for the same muscle group.

Elastic stretching is used here to denote the activity of rhythmically swinging an arm or leg towards its most outer position. This swinging action has nothing to do with stretching the muscles. Elastic stretching has been used, and should be used, when warming and limbering up.

On the other hand, stretching exercises are those designed to lengthen the muscle groups involved quickly, which in turn, increases the range of motion in the joints. Thus, both stretching exercises and elastic stretching exercises are used to promote general agility.

Assume that a person wants to increase the range of movement of his arm so that he can draw it back as far as possible (swimmer, thrower, gymnast).

Fig. 47.

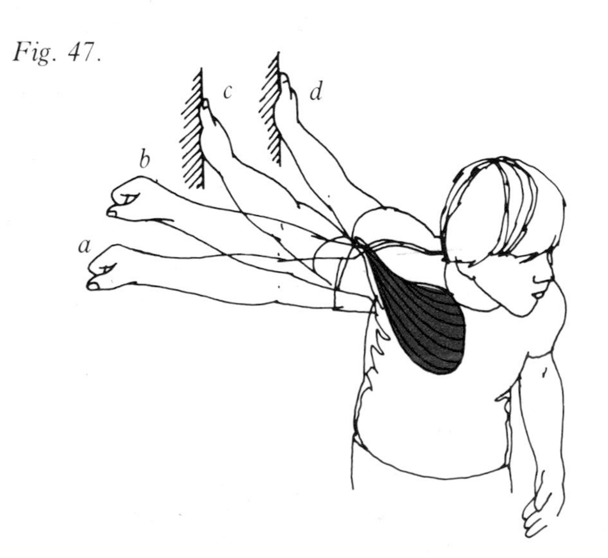

(a) If the arm is swiftly swung back, it could perhaps reach position *a*. What happens is that the muscles situated behind the shoulder give the arm its speed, but the arm is halted before it reaches an outer position. This halting is caused by the spindles in the muscles at the front of the joint and endings in the joint itself which send warning signals to the CNS. The CNS, in turn, sends contraction signals to the same muscles in order to prevent injury of the cells. Instead of stretching, the muscles on the fore side of the joint are forced to contract to protect themselves. This type of elastic stretching may produce eccentric strength but it certainly does not train flexibility. When elastic stretching is performed forcefully by a person who is not warmed-up, the risk of minor ruptures is obvious. However, easy elastic stretching is a good way of toning the muscle.

(b) If, instead, the arm is slowly drawn back as far as it will go, the position it reaches will generally be farther than that of the swinging arm described above. Just how far back depends on the suppleness of the muscles at the front of the joint, and the strength of the muscles which pull from the back of the joint. This method is called active stretching. Work is actively being done by the antagonists (the muscles on the other side of the joint) to the muscles we want to stretch. However, we try as far as possible to relax the actual muscles that are to be stretched.

(c) In passive stretching the arm is pressed backwards further with the help of external forces (i.e. not the muscles at the back of the shoulder). This could be done, for example, with the aid of a friend, or the hand could be held against a wall and pressed, thereby utilising the force of the legs. Passive stretching is always more effective than active stretching in reaching the outermost position.

(d) The best results will be obtained, however, by first extending the arm as far as possible by the passive stretching method. (The muscle group that is to be stretched should be as relaxed as possible.) Next, try to contract the muscle group for a few seconds (6 s) while the external force (friend, legs, wall) prevents any movement occurring at the joint. The muscle tension is thus static, which implies that the muscle belly is somewhat shortened and the collagenous fibres of the tendons are somewhat extended.

Fig. 48.

By relaxing again (2–4 s), after which passive stretching is again exercised, the arm could reach even farther. The method of contracting a muscle with the aim of increasing its ability to stretch, is called the PNF-method (Proprioceptive Neuromuscular Facilitating). The PNF-method is used by physiotherapists when they train an injured muscle in an effort to restore its natural length. It is also called the contraction–relaxation–stretching method.

Contraction affects the tendon organs of the muscle so that they send out inhibitory signals to the muscles during the ensuing relaxation. The limb should be close to the outermost position the whole time, and any movements should be slow and easy so that the muscle spindles are prevented from sending contraction signals to the muscle.

Fig. 49.

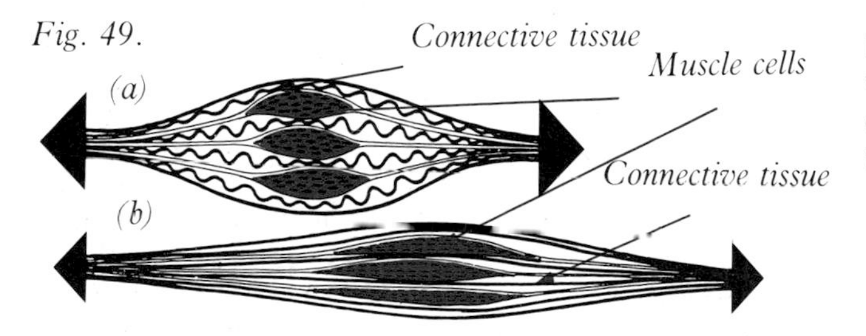

Figure 49 (a) shows what happens when one tries to stretch a contracted muscle. The external force acting on it affects only the muscle cells, and consequently only minor muscle ruptures can result.

Figure 49 (b) shows that when the muscle is relaxed, it is the fibres of the connective tissue that are stretched. Connective tissue is rigid and can resist swift stresses (jerks, elastic stretching) but yields to prolonged stress (remain in the outermost position for about 30 s).

Examples of flexibility training

The example below is designed to demonstrate further how a muscle group can be stretched effectively, i.e. the flexibility of a particular joint improved. Good flexibility in the hip joint is of great importance to nearly all athletes. The group of muscles that are easily injured and which impede effective movement patterns, are the muscles of the groin (the adductors). These originate from the pubic bone and are inserted into the thigh bone (see p.41). When sitting on the floor with the soles of the

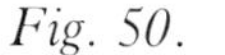

Fig. 50.

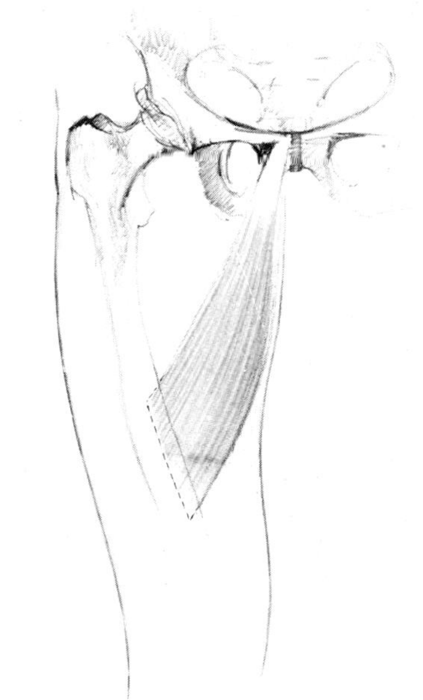

feet together and as close as possible to the body, it will be seen that the length of the groin muscles will determine just how far the knees can be pressed to the floor.

Fig. 51.

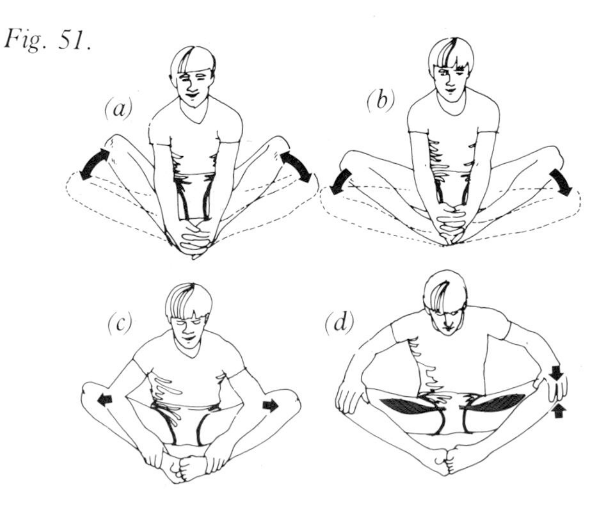

When, in the same position, the knees are bounced up and down, the activity corresponds to that described in (a) on p.24. Further, if the knees are pulled down using the muscles of the outer side of the hip (abductors, p.39), the activity corresponds to the method described in (b) on p.24, i.e. active stretching of the groin muscles. Pressing the knees down with the hands corresponds to the method described in (c) on p.24, i.e. passive stretching of the groin muscles. The most effective method, i.e. the PNF-method (method (d) on p.24), is practised in the following way.

Press down with your hands holding back with your knees (approx. 6 s), relax (2 s), press down with your hands again (10 s), hold back once again (6 s), relax (2 s) and finally press with your hands for about 10 s with relaxed groin muscles.

The straight knee extensor muscle (rectus femoris, p.24) can be stretched using the PNF-method in the following way. Take your foot in your hand. Bend your knee maximally and try to keep your hip as straight as possible. Next, try to extend the knee joint holding back with your hand for about 6 s (the thigh muscle thus contracts statically).

Fig. 52.

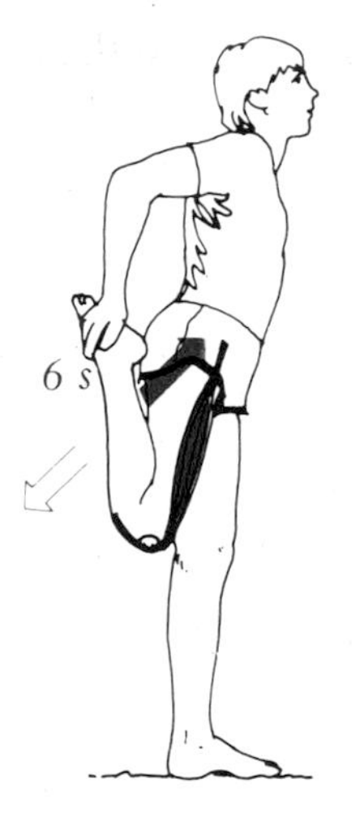

Next, relax for 2 s. Then slowly press your hip forward a little and stand still for about 10 s. Repeat this several times.

It has been shown that the number of muscle injuries in, for example, football has been radically reduced when flexibility exercises were introduced as a regular part of the training programme. A properly executed stretching exercise following the PNF-method, takes about 20 s. A ball player should perhaps work through five different muscle groups about three times on each training occasion in order to obtain good results: 20 x 5 x 3 = 300 s. Thus, no more than 5 min are required each time he trains. Many trainers claim that they have no time for flexibility training; but better knowledge of how to perform this type of training, as well as its benefits, should dispel such arguments. However, a prerequisite for this is that the trainer acquires in-depth knowledge of the origin, insertion and function of muscles. Suitable positions for stretching exercises are shown with the aid of diagrams after each part of the body has first been described anatomically and also in Chapter 7.

E. Strength Training

It is customary to distinguish between different types of strength.

Fig. 53.

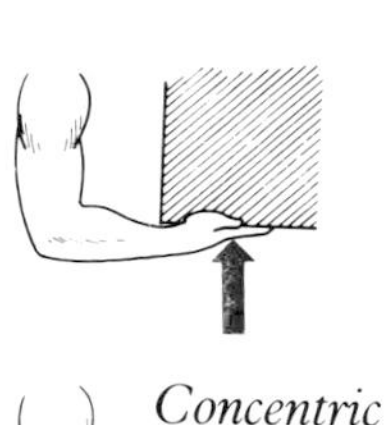

Static maximum strength
Static endurance

Concentric

Dynamic maximum strength
Dynamic endurance

Eccentric

In addition, dynamic strength is divided into concentric and eccentric strength (p.19).

The majority of cells in the body are capable of reproducing themselves. This, however, does not apply to the striated muscle cells to any great extent. Their number is largely determined by genes. When training strength, the muscle cells are thus not increased. Instead, there is an increase in the number of myosin and actin filaments, which are responsible for muscle contraction. A certain division of cells takes place. Like the majority of cell types, the muscle cells follow the principle of overcompensation. According to this principle, training breaks down parts of the stressed structures. The body compensates for this by producing new fibrils, the number of which is somewhat greater than the number of broken down fibrils. The athlete should begin his next training session at the time when overcompensation has reached a maximum.

Fig. 54.

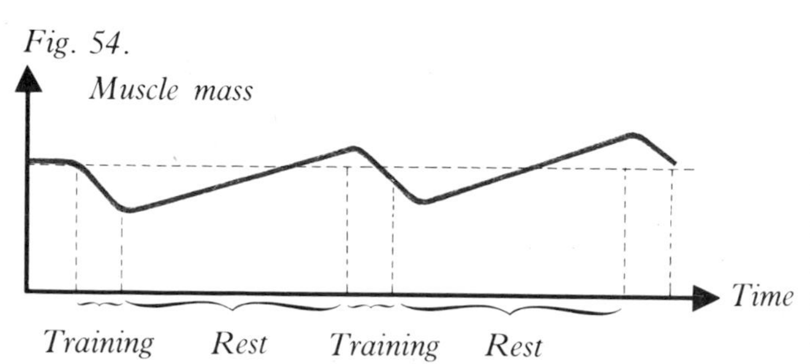

When the training sessions are too close together, or the training too vigorous without extra long pauses, the regeneration of fibrils never reaches the previous level. The athlete thus slowly wears down his body; he "becomes overtrained".

As a general rule we can say that recovery normally takes about 24–48 h. (After extremely vigorous training, about 72 h.) Thus, an athlete should train relatively vigorously three times a week in order to increase his strength quickly.

Strength decline is first noticeable after 5–6 days without training. About one session a week should therefore suffice for an athlete to maintain the strength he already has.

As mentioned earlier, youths should not train with loads exceeding their own bodyweight before or during puberty. Because children younger than 11 or 12 years find complicated movements difficult, they should avoid exercises demanding coordination. Training with light weights aims at accustoming young people to the equipment as well as training the correct technique.

On the basis of what has been said about motor units (p.23), strength should be trained by using heavy work loads (all the motor units must be activated). A common target for an athlete is 80–95% of what he can maximally manage. In this way he improves his ability to work at a dynamic maximum. If he wishes to improve his dynamic endurance, he should work with loads that correspond to 20–25% of his maximal capacity. Endurance training has the effect of decreasing both the maximal strength and speed of the muscle group in question. Training for speed is undertaken with a work load of 50–80% of the maximum that can be managed by the athlete.

There are two main types of muscle cell: white and red (see p.23). They have received their names from the colour they assume when they are specially treated to be examined under the microscope. The fast muscle cells (type II) described on p.23, are the white cells, and the slow muscle cells (type I) are the red. White muscle cells contain many actin and myosin filaments (see p.15). Red muscle cells contain relatively few but, on the other hand, have more energy-supplying components.

These conditions result in strong, fast white cells that tire quickly; and red cells that are characterised first and foremost by their endurance. Both red and white fibres are contained in the same muscle. A muscle's composition is determined by genes, but it can, in part, "be changed" by specific training.

Different sporting events require different types of strength, so athletes and trainers should choose exercises with care. Weight-lifting is dependent on strong, fast muscles. Weight-lifters training should, therefore, involve few exercises that are designed for dynamic endurance. It has been shown that pure endurance training considerably reduces both the strength and speed of the trained muscles.

The concepts "set" and "repetitions" (reps) are often used in descriptions of strength training. A set contains a particular exercise that is repeated a certain number of times. A set can, for example, consist of six consecutive repetitions. (Various scientific investigations indicate that the quickest increase in strength is acquired with 6 reps.) Several reps with a light work load promotes endurance. Fewer reps with a heavier work load develops maximal strength.

If we take, as an example, an exercise that strengthens the pectoralis major (the largest chest muscle), then a weight should be chosen that can be lifted and lowered six times with slightly bent arms (to avoid straining the elbow joint). This should be followed by a period of rest (2–3 min). During the pause, a completely different muscle group (e.g. the knee extensors) can be trained in the same way. The exercise for the chest muscle is then repeated, i.e. a second set with its 6 reps. An instruction which reads "5 sets x 6 reps" means a total of 30 lifts with five 2–3 min pauses after each set.

Fig. 55.

The work load can be varied as the amount of arm flexion varies. (See the mechanics section on levers, p.34.) A muscle is stronger when it works eccentrically than when it works concentrically. This is easily verified by slowly lifting a weight; it feels easier to lower than to lift. If the aim is to train eccentric strength, the athlete should keep his arms almost straight while lowering weights, and bend them while lifting.

A curve which presents an approximation of the number of times a movement can be performed has been developed experimentally. This number depends on the ratio of work load to the maximal load the muscle group in question can manage.

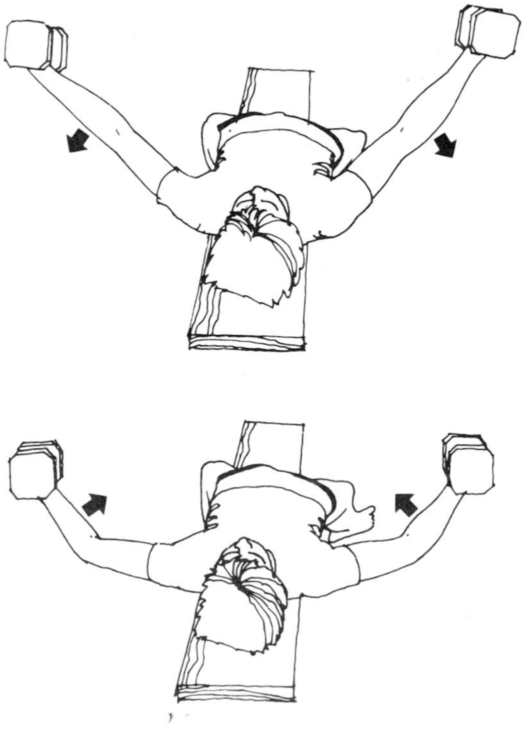

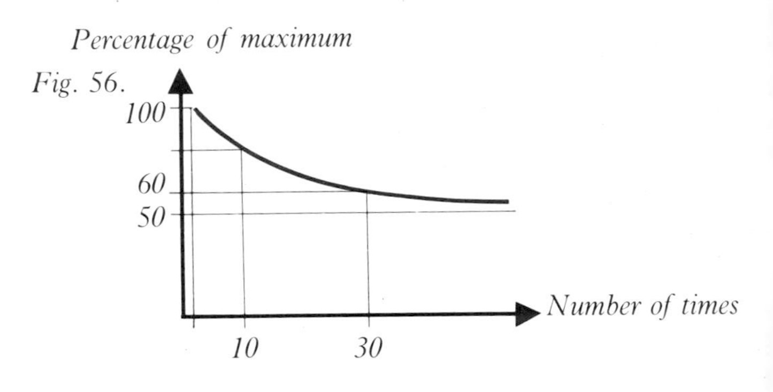

Fig. 56.

Another curve (Rhomerts curve) shows how many minutes an isometric contraction can take if it falls short of the isometric maximum of a muscle group.

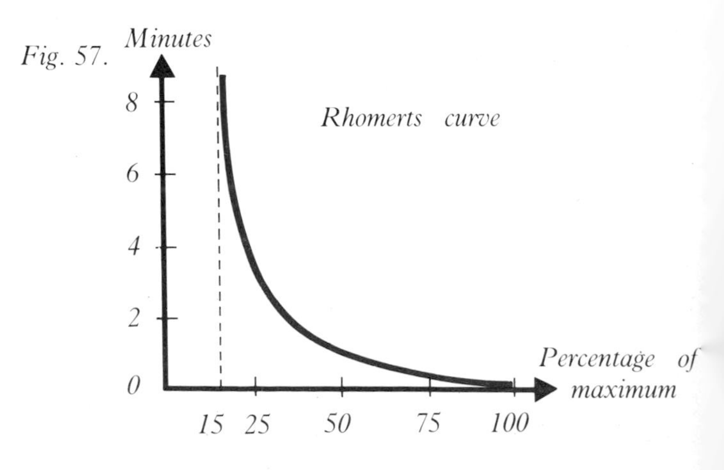

Fig. 57.

A method for quickly building up the strength of muscles is the so-called pyramid system. For each set the work load (weights) is increased and the number of reps is decreased so that the effort exerted in the final set is maximal.

Set 1: 6 reps with 70% of the maximum weight that can be lifted 6 times.

Set 2: 5 reps with 75% of the maximum weight that can be lifted 5 times.

Set 3: 4 reps with 80% of the maximum weight that can be lifted 4 times.

Set 4: 3 reps with 85% of the maximum weight that can be lifted 3 times.

Set 5: 2 reps with 90% of the maximum weight that can be lifted twice.

Set 6: 1 rep with 95–100% of the maximum weight that can be lifted once.

It should be pointed out that the strength of muscles increases faster than the strength of tendons, ligaments and cartilage. Extremely vigorous training can, therefore, injure muscle attachments and joints.

The table below give approximate work loads together with the number of times exercises should be carried out in order to develop different types of strength. Exercises for developing strength are described in each section of the book where the anatomy of different parts of the body is dealt with (e.g. the knee joint, p.50).

	Endurance	*Speed*	*Maximal strength*
% of max.	*25–50*	*50–80*	*80–100*
No. of reps	*more than 40*	*approx. 10*	*1–6*
No. of sets	*5*	*4*	*3*

Chapter 2

BASIC RULES OF MECHANICS

In order to understand how the skeleton is constructed and how the muscles affect a particular part of the body, we must familiarise ourselves with certain properties of force as well as what is meant by the concept "moment of force".

A force (F) is represented by an arrow which indicates its size and direction.

Fig. 58.

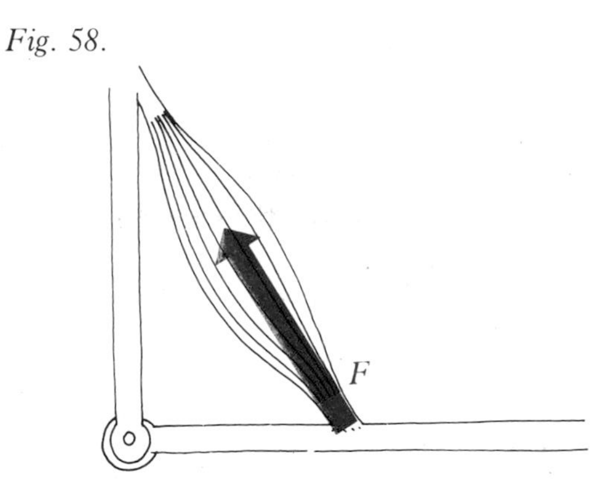

If an object weighs 5 kg, it is said that the gravitational force acting on it is 50 Newtons (50 N).

Fig. 59.

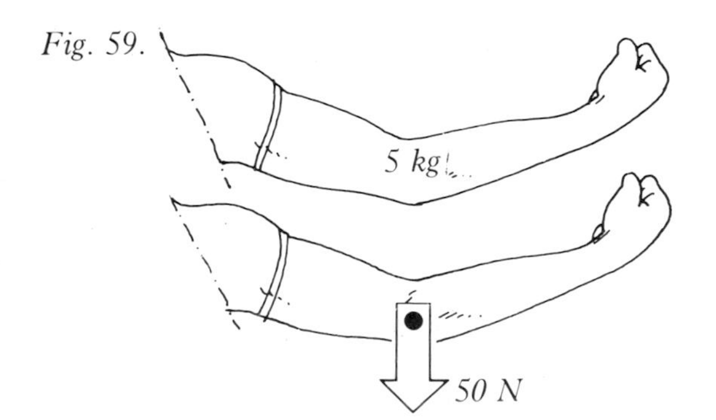

Fig. 60.

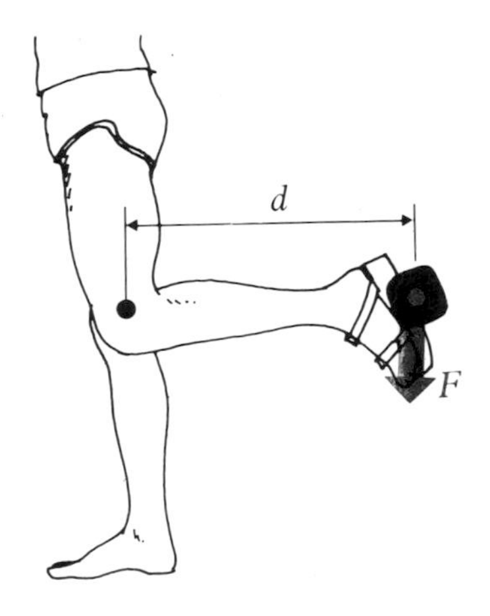

If a force is applied to a body at a certain distance *(d)* from a given point, it may happen that the force causes a certain angular motion at that point. This tendency toward angular motion is called the moment of force and is represented by M. The moment of force is calculated by multiplying the magnitude of the force by the length of the distance *(d)* between the force and the point.

Thus, $M = F \times d$.

The magnitude of the moment of force tells us just how much the object is striving to turn. Thus, the moment of force is a number which depends on two factors at the same time; in particular, the same force can have different turning effects depending on its point of application.

In the examples below, the force is the same but the moment of force varies such that it is least in example 1, and most in example 3. Suppose that the force is 50 N and the lengths are: $d_1 = 0.20$ m, $d_2 = 0.30$ m and $d_3 = 0.50$ m. The moment of force is then, according to the formula $M = F \times d$.

Fig. 61.

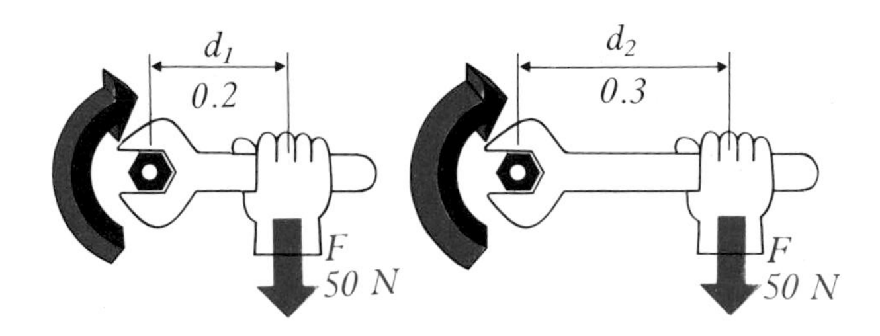

$M_1 = 50$ N x 0.20 m
= 10 N m

$M_2 = 50$ N x 0.30 m
= 15 N m

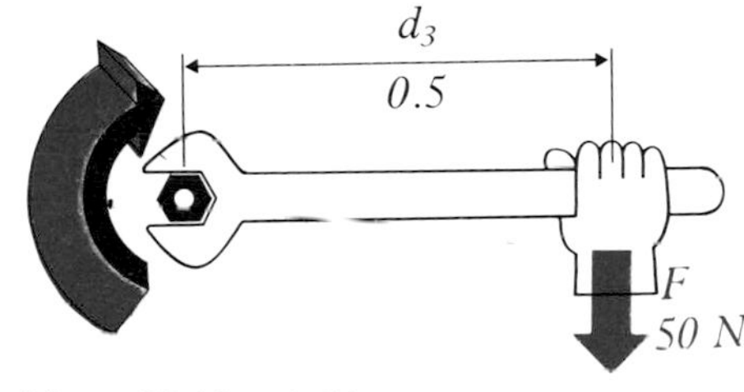

$M_3 = 50$ N x 0.50 m
= 25 N m

The numbers 10, 15 and 25 show how much the wrench handle tends to turn. The moment of force of 25 Nm can be obtained in many different ways. An alternative to example 3 would be to apply a force of 100 N at a distance of 0.25 m from the bolt. $M_4 = 100 \times 0.25 = 25$ Nm.

Fig. 62.

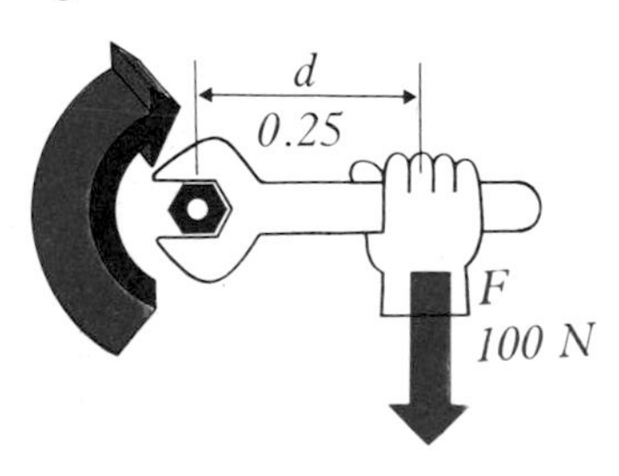

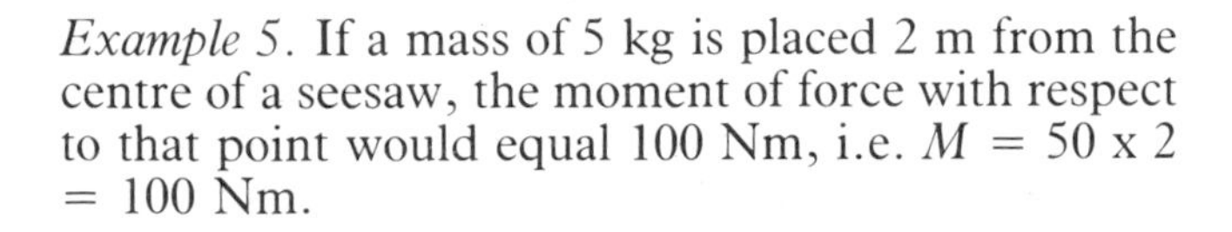
Example 5. If a mass of 5 kg is placed 2 m from the centre of a seesaw, the moment of force with respect to that point would equal 100 Nm, i.e. $M = 50 \times 2$ = 100 Nm.

Fig. 63.

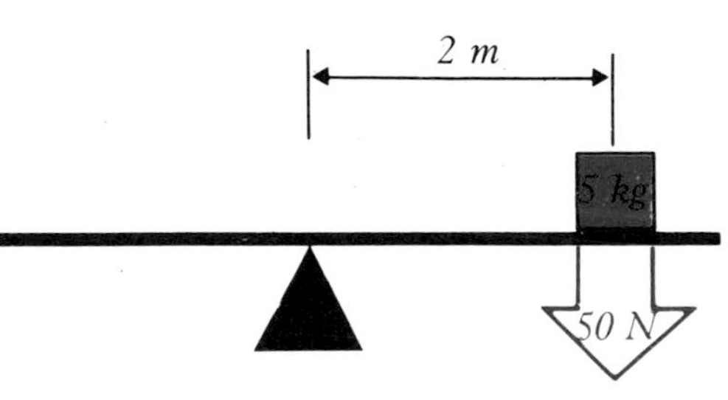

Example 6. The moment of force of 100 Nm can also be achieved by placing a 10 kg load 1 m from the centre.
$M - 100 \times 1 = 100$ Nm.

Fig. 64. Example 6.

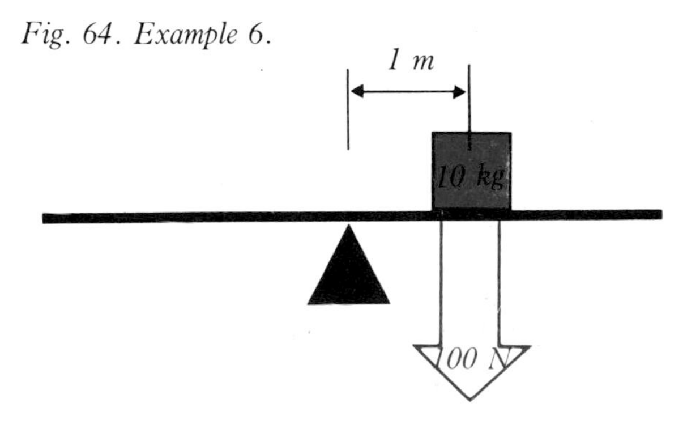

Example 7. If weights are placed on each side of the seesaw's centre as in example 7, then the seesaw is in equilibrium. It is said that the moments cancel out (seesaw principle). 100 x 1 = 50 x 2.

Similar reasoning can be applied to the way in which muscles affect different parts of the body.

Fig. 65. Example 7.

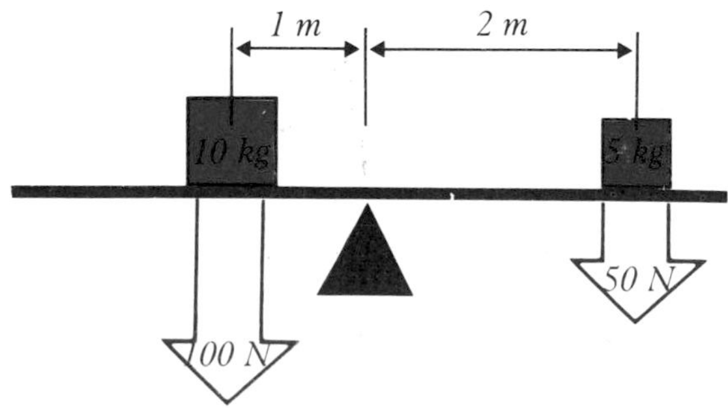

Example 8.
F_m = Muscle force.
d_m = The muscle's lever arm to the centre of the joint.
F = The force of gravity acting on the ball.
d = Distance to the centre of the joint.

Fig. 66. Example 8.

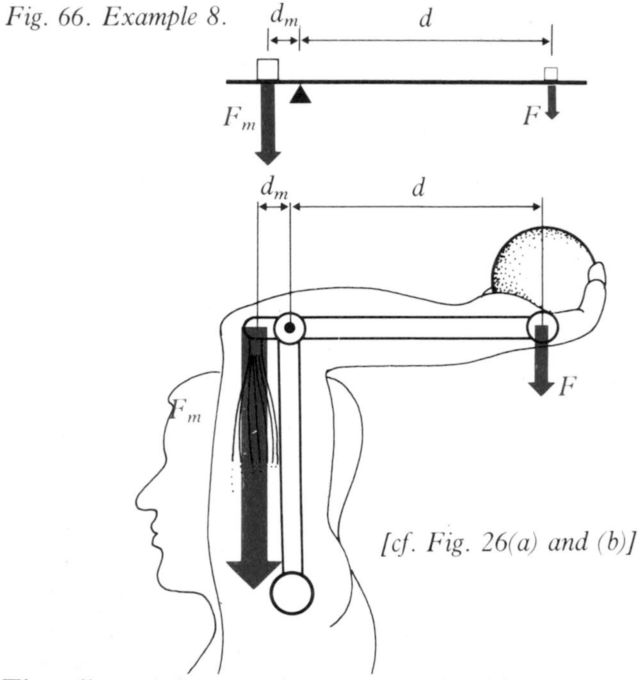

The elbow joint can be compared with a seesaw where the external force (F) acts on one side and the internal force (F_m) acts on the other. The arm is held still if $F_m \times d_m = F \times d$.

Example 9. The body is in equilibrium when $F_m \times d_m = F \times d$. If $F_m \times d_m$ is greater than $F \times d$, the body accelerates upwards (rises even further on its toes). If $F_m \times d_m$ is less than $F \times d$, the body will not be able to hold its position and will sink. The numbers refer to a person of mass 80 kg.

Fig. 67. Example 9.

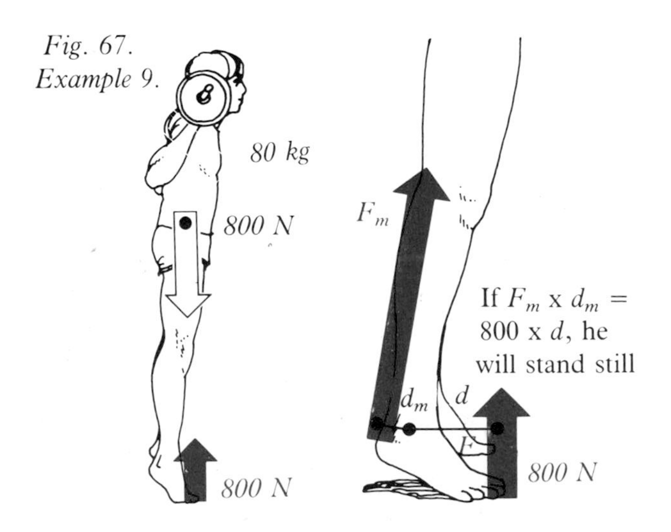

Example 10. The weight of the trunk (30 kg) is counterbalanced by the traction force of the back muscles. If the trunk's weight lies three times as far in front of the vertebrae as the back muscles lie behind them, then the traction force of the back muscles must be three times greater than the trunk's weight.

Fig. 68. Example 10.

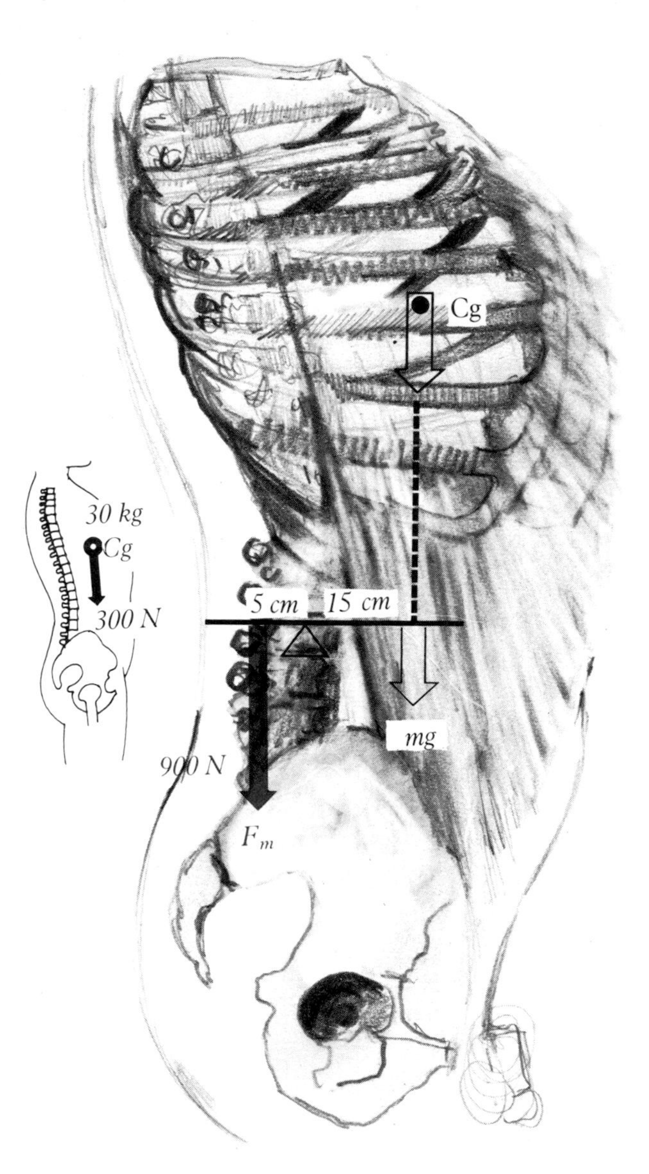

In general, the part of the body in question will accelerate in the direction where the moment of force is greater.

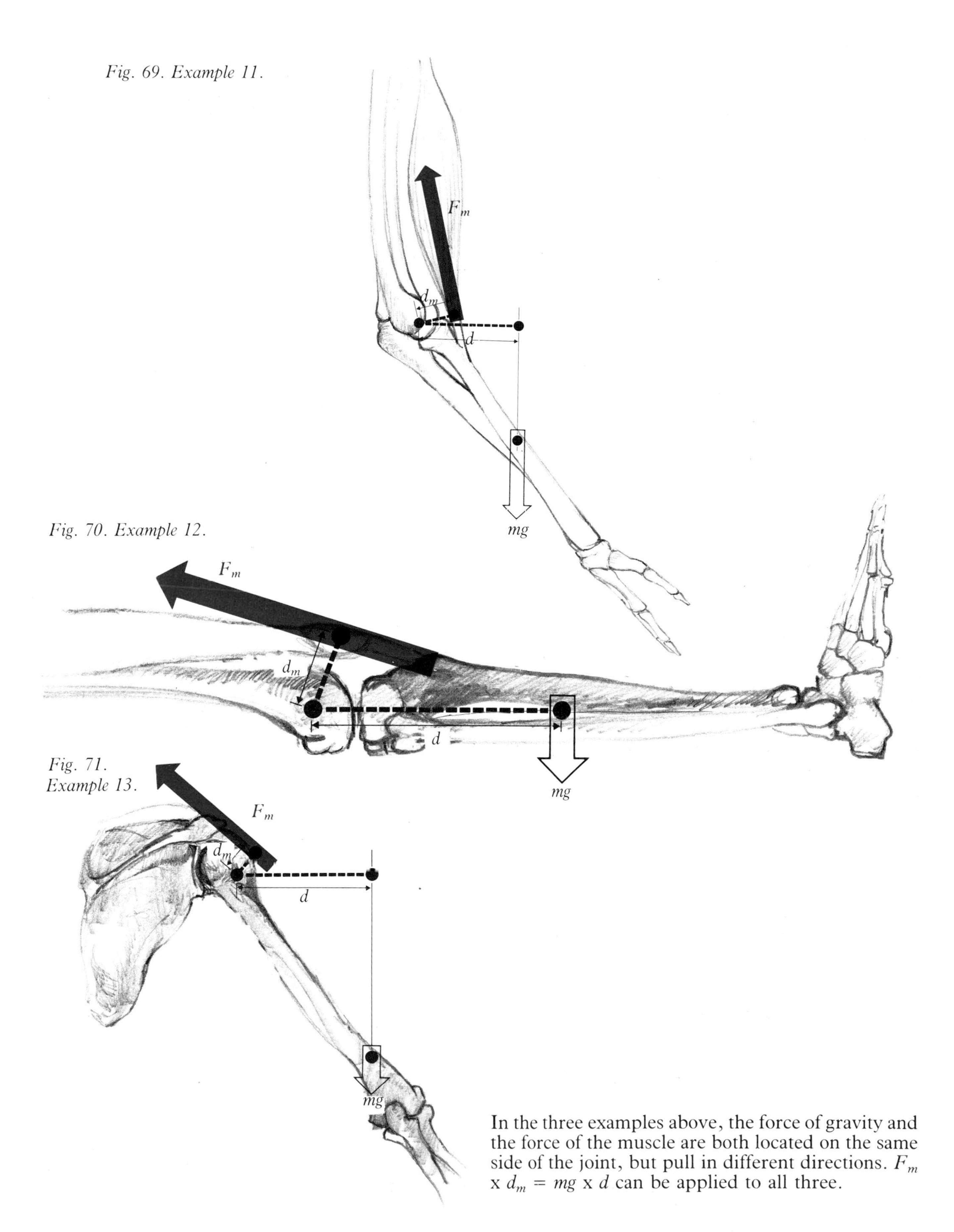

Fig. 69. Example 11.

Fig. 70. Example 12.

Fig. 71. Example 13.

In the three examples above, the force of gravity and the force of the muscle are both located on the same side of the joint, but pull in different directions. $F_m \times d_m = mg \times d$ can be applied to all three.

Fig. 72: Example 14.

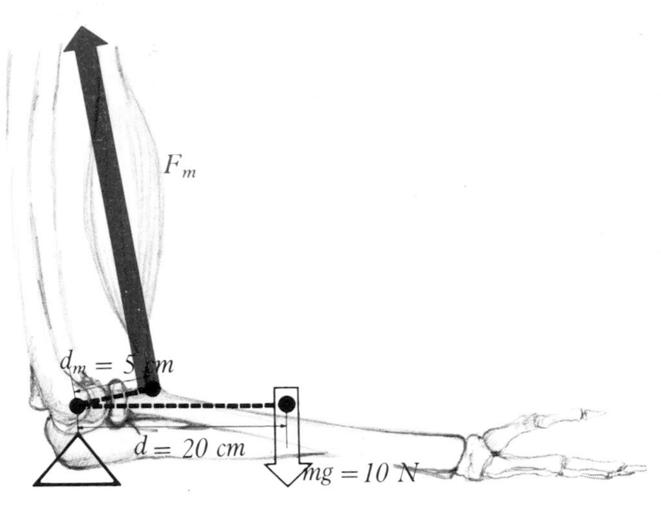

The forearm (1 kg) is acted on by a gravitational force of 10 N. If the centre of gravity lies 20 cm from the joint, then the moment of force will be equal to 10 x 20 Ncm. If the "muscle" that bends the elbow is attached at a point 5 cm from the joint, then F_m x 5 must equal 10 x 20 if the arm is to be held still.

The diagrams illustrate different types of joints and the way in which the body has solved the problem of resisting (or creating) movements.

If the point of insertion of the arm's most important flexor differs in two people (A and B), then these two people will differ greatly in the strength and speed of their arms.

Example 15. Suppose the muscle of A is attached at a point which gives a lever arm of 4 cm when the elbow is bent at a 90° angle.

Fig. 73. Example 15.

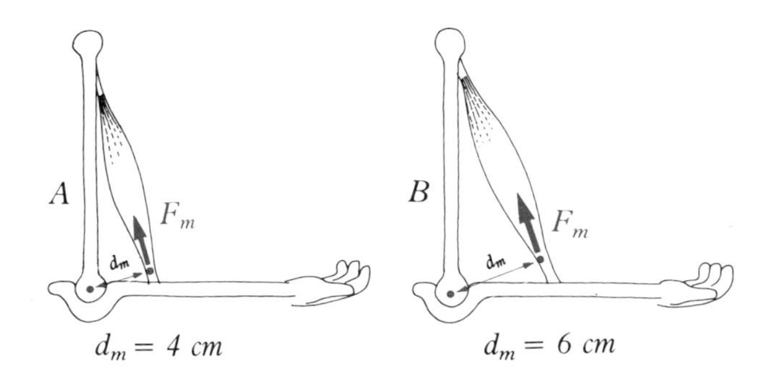

The muscle of B is attached at a point further from the joint, which means that the perpendicular distance to the muscle force is, say, 6 cm when the elbow is bent at a 90° angle.

If the muscle force (F_m) is equal for both people, the moment of force (strength) will be 50% greater for B than for A. Thus, B is stronger than A.

If the muscle is shortened by 3 cm in both cases, the forearm of A will move through a greater angle than the forearm of B. Thus, A is faster than B (Figure 74).

Fig. 74.

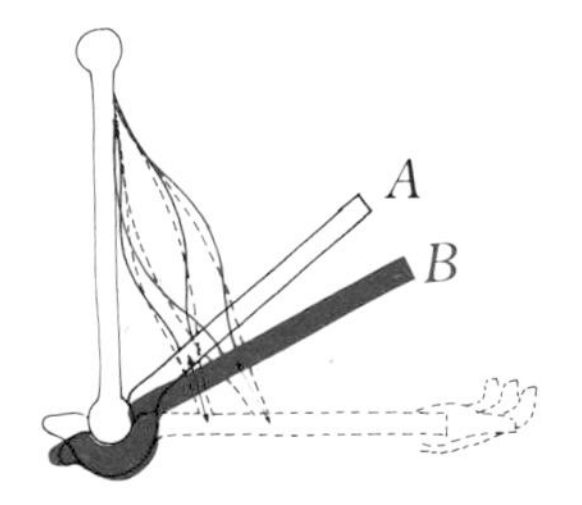

The ability of a muscle to develop force (F_m) can be influenced by strength training, but a muscle's attachments cannot be changed. In other words, we are constructed in such a way that we are particularly suited to certain activities and less suited to others — at least so far as top performance is concerned.

Tendon and ligament forces are passive (1; Figure 75). This means that they are brought about by external forces (2; Figure 76) or the muscles (3; Figure 77).

If we consider a small part of a taut tendon, we can understand that it strives to stretch. Thus, we can draw an arrow indicating force either in the direction of muscle pull or in the direction of tendon (muscle attachment) resistance (1).

Suppose a person is standing with knees slightly bent and that one of his knee extensors is contracted with a force of 1000 N. The force acting on its origin is then 1000 N (a) and is directed toward the knee. A force of 1000 N (c) is acting on the kneecap and is directed toward the thigh. A bundle of connective tissue somewhere in the muscle stretches with a force of 1000 N (b). The tension in the tendon extending between the kneecap and the shin bone is also about 1000 N. The force acting on the apex of the kneecap, the tendon's attachment and a point on the tendon is, according to Figure 79, 1000 N.

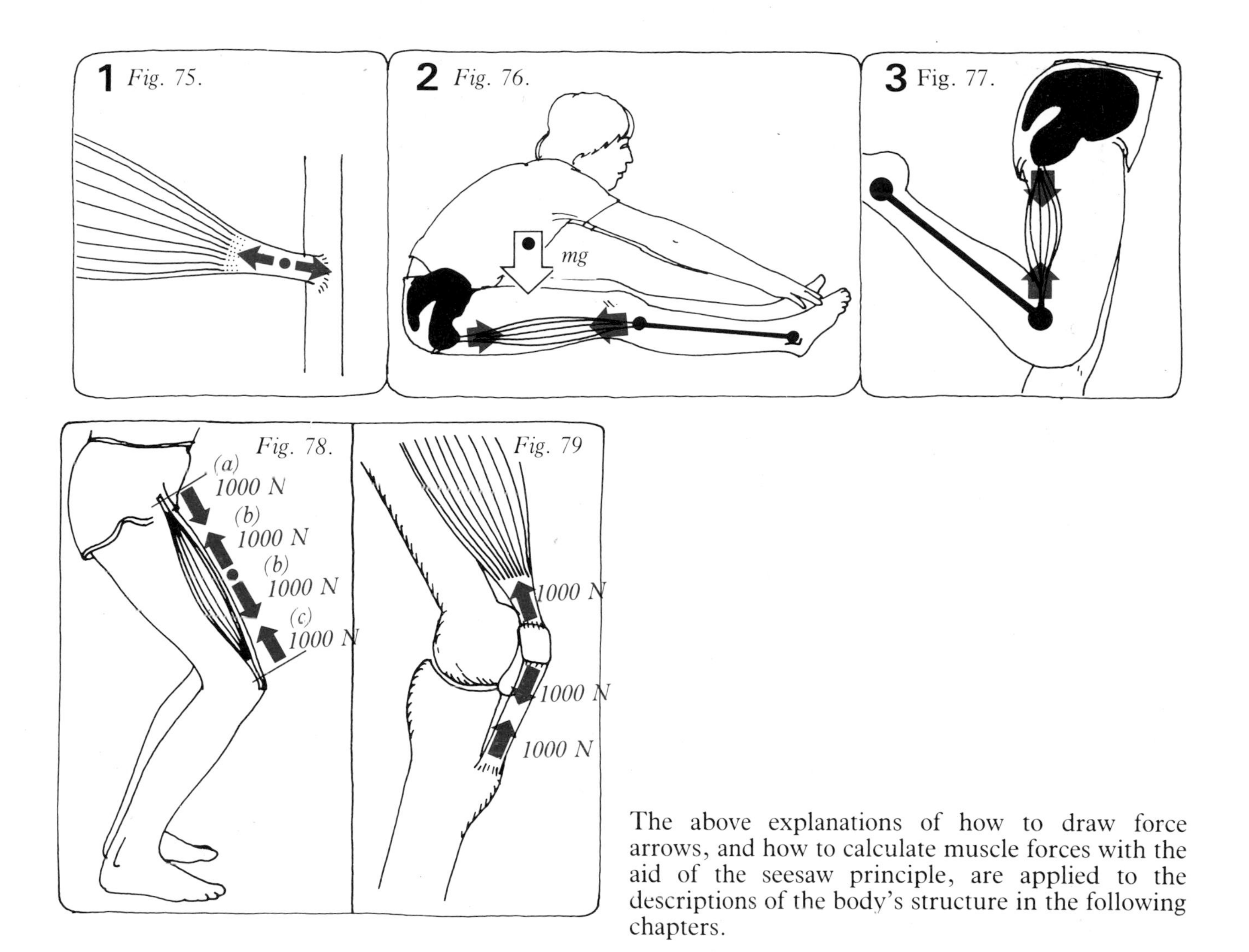

The above explanations of how to draw force arrows, and how to calculate muscle forces with the aid of the seesaw principle, are applied to the descriptions of the body's structure in the following chapters.

Chapter 3

ANATOMY AND FUNCTION OF THE LEG

In order to analyse different types of movement correctly, we must first study the anatomy of the area concerned. Here, we will examine the origin, insertion and function of the larger muscles. In addition, we will look at the joints' potential for movement and their limitations.

A. The Hip

The majority of muscles responsible for moving the hip joint originate from the pelvis; some originate from the spinal column. Some of them also pass over the knee joint. Thus, we must acquaint ourselves with the parts of the skeleton shown in Fig. 80.

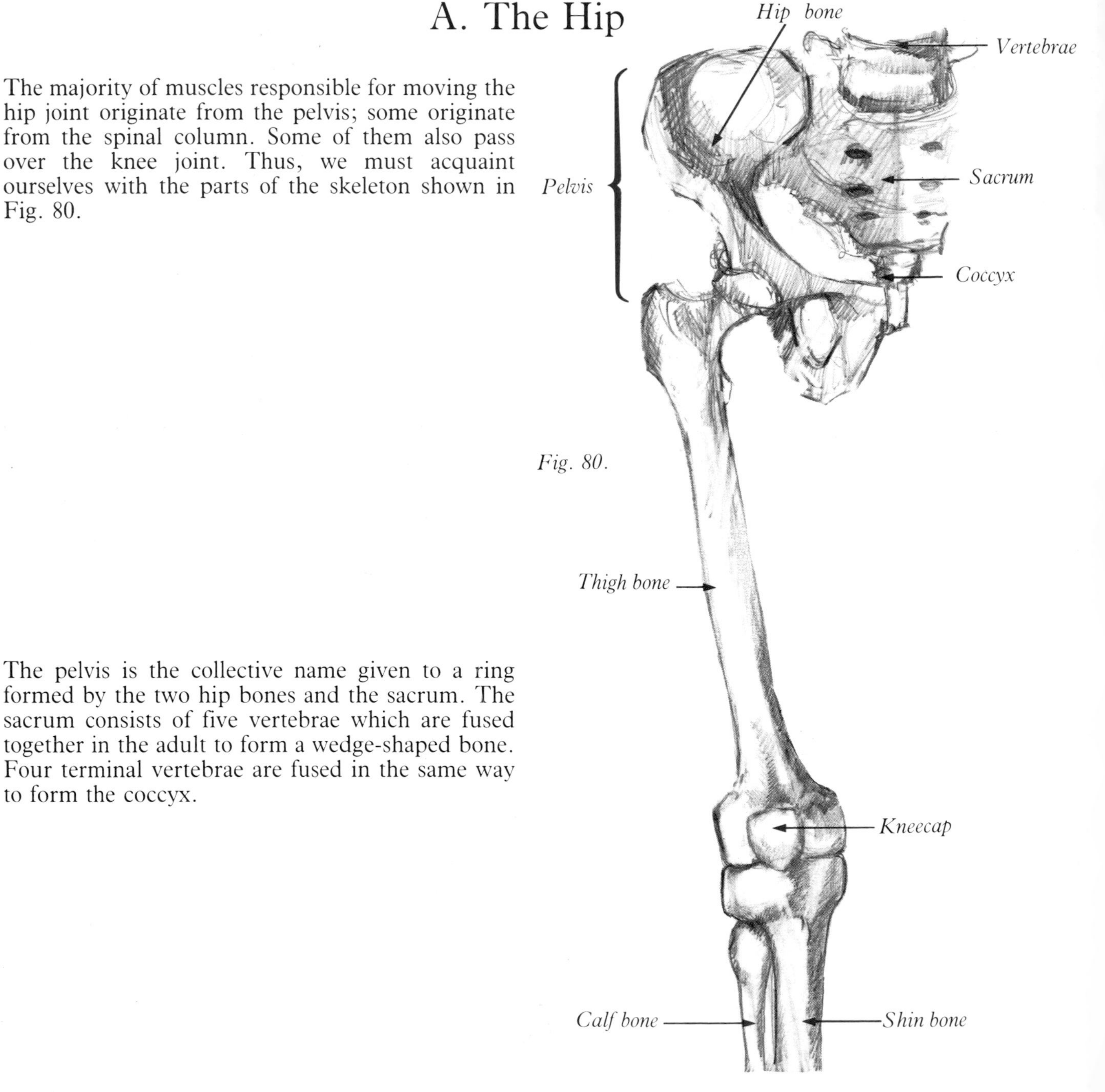

Fig. 80.

The pelvis is the collective name given to a ring formed by the two hip bones and the sacrum. The sacrum consists of five vertebrae which are fused together in the adult to form a wedge-shaped bone. Four terminal vertebrae are fused in the same way to form the coccyx.

Fig. 81.

e
a
b
Ilium
c
Pubic bone
Ischium
d

Right hip bone, external view.

The hip bone has developed from three separate centres of ossification, which is why the following distinction is made.

1. Ilium.
2. Ischium.
3. Pubis.

All the cavities, outgrowths and spines have different names. Those that are directly connected with important muscles are given here.

(a) Anterior superior iliac spine.
(b) Anterior inferior iliac spine.
(c) Acetabulum.
(d) Ischial tuberosity.
(e) Iliac crest.

The hip joint is a so-called ball and socket joint which means that it can move in all directions (p.13). There are certain extra-capsular structures that give strength to the joint and, in particular, prevent the leg from swinging outwards and backwards. Backward swinging is impeded by the powerful ligament which is attached to the iliac part of the hip bone and which passes downward to the thigh bone (a) iliofemoral ligament. Outward swinging is restricted by (b) pubofemoral ligament.

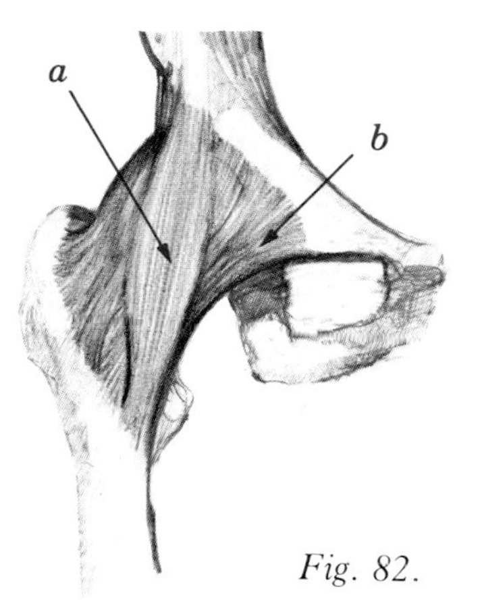

Fig. 82.

The diagrams below demonstrate actions that demand great flexibility of the hip.

It has not been shown that any injury is sustained when a person stretches in the directions mentioned above. If, on the other hand, a person compensates for a poor ability to swing his leg backwards by overstretching the lumbar region of his back, he would in probability be literally "making a rod for his own back". Overstretching the lumbar area often leads to pain of a more or less serious character.

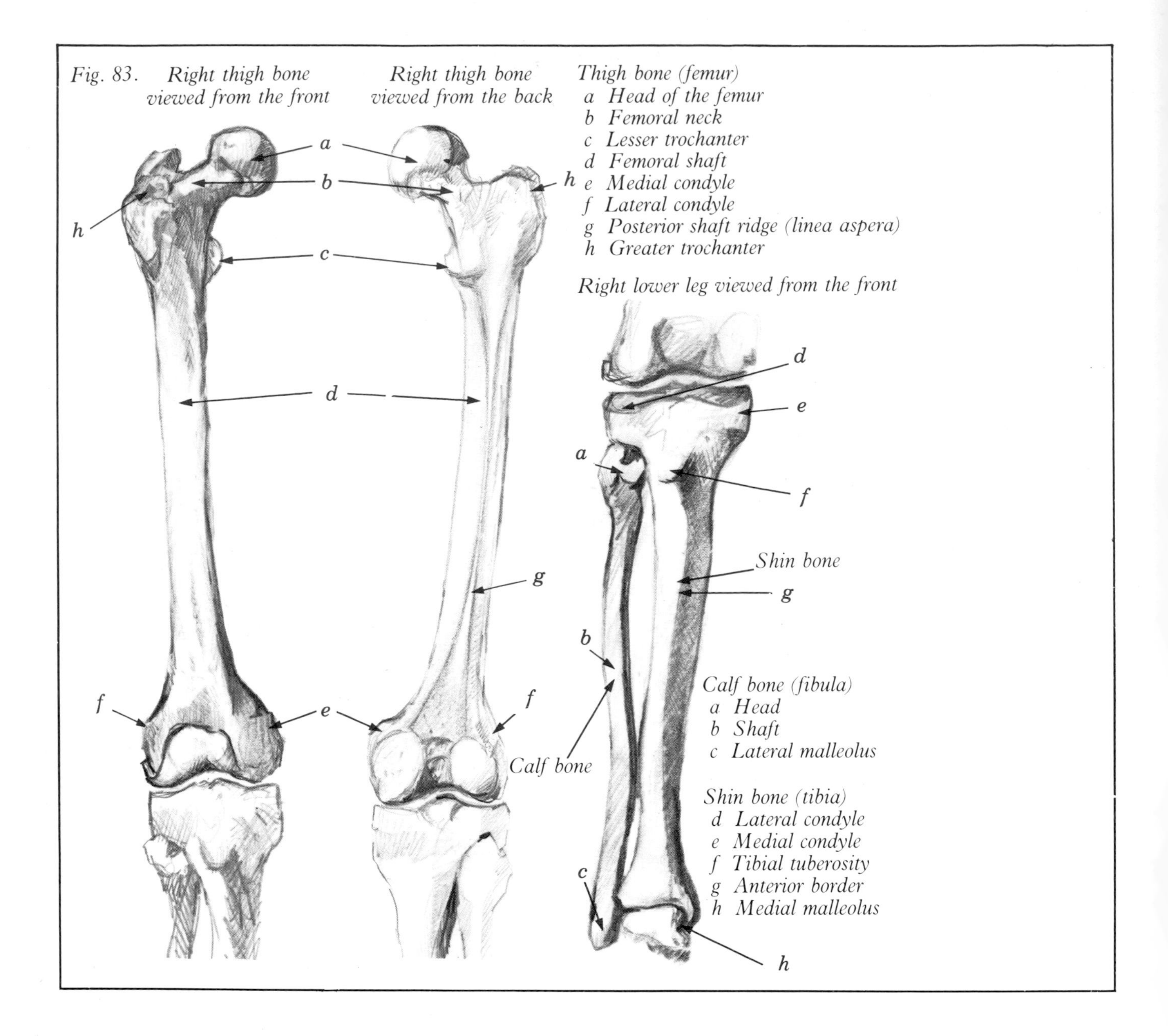

Fig. 83.

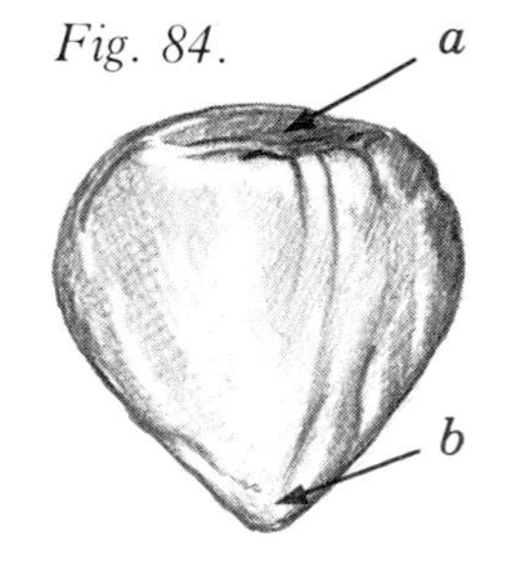

Fig. 84.

The kneecap (patella) has (a) a base that is directed upwards and (b) an apex that points downwards. The inside of the kneecap is covered with a 6-7 mm thick layer of cartilage whose surface articulates with the cartilage-covered condyles of the thigh bone.

B. Hip Muscles

The following are the most important muscles that pass over the hip joint.

1. Buttock muscles.
2. Groin muscles.
3. Hip flexors.

1. Buttock muscles

(a) Large buttock muscle (gluteus maximus).
(b) Intermediate buttock muscle (gluteus medius).
(c) Small buttock muscle (gluteus minimus).

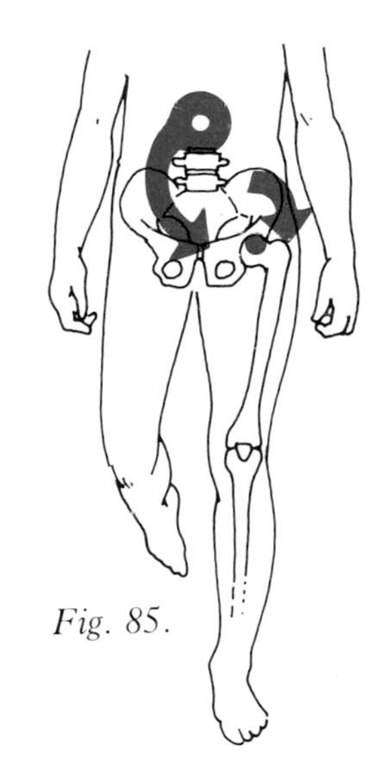

Fig. 85.

Two of the three muscles are attached to the greater trochanter. The intermediate and small buttock muscles have such a large area of origin that they can move the thigh bone in all directions except inward towards the midline of the body (adduction). These muscles are activated by walking and running. They have the important task of stabilising the hip joint when the corresponding foot only is in contact with the ground. This stabilisation is necessary in order to prevent the upper body from falling to the opposite side.

The buttock muscles are subjected to great stress when a person runs uphill (the large buttock muscle is responsible for the powerful backward drive of the leg) or downhill (medius and minimus stabilise the hip). The medius and minimus then work

Fig. 86.
(a) Large buttock muscle (gluteus maximus).

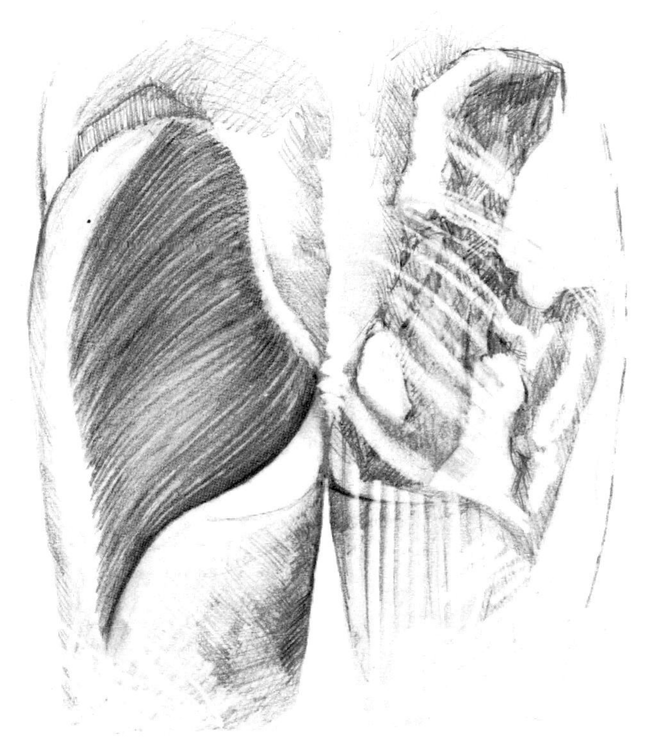

Fig. 87.
(b) Intermediate buttock muscle (gluteus medius).

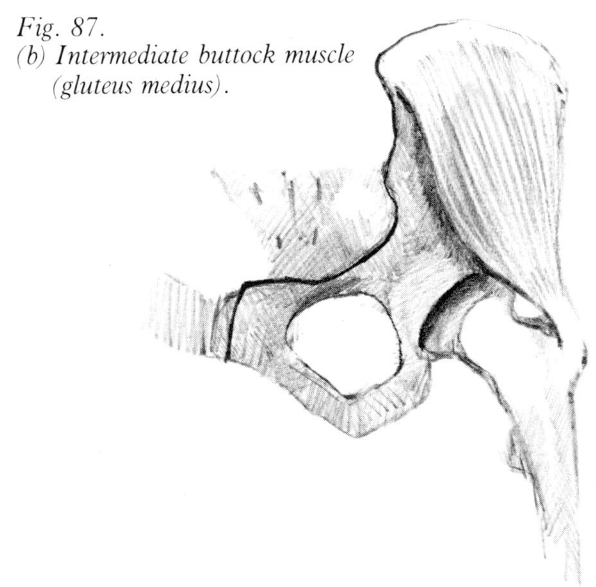

Fig. 88.
(c) Small buttock muscle (gluteus minimus).

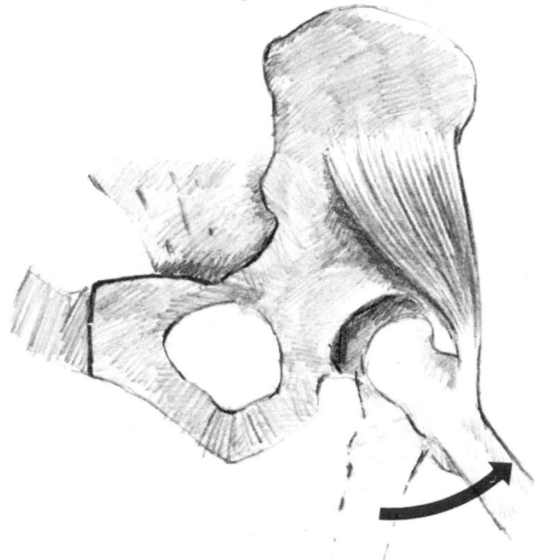

eccentrically, i.e. they restrain the upper body so that it does not fold inwards at every step (Figure 85).

Examples of exercises for the buttock muscles: running; jumping on one leg; and, while standing on one leg, lowering and raising the opposite side of the pelvis. Further exercises for the medius and minimus are: (a) lying on one side and lifting the upper leg and (b) lifting the upper part of the body as far as possible while the legs are fixed (Figure 89).

Fig. 89.

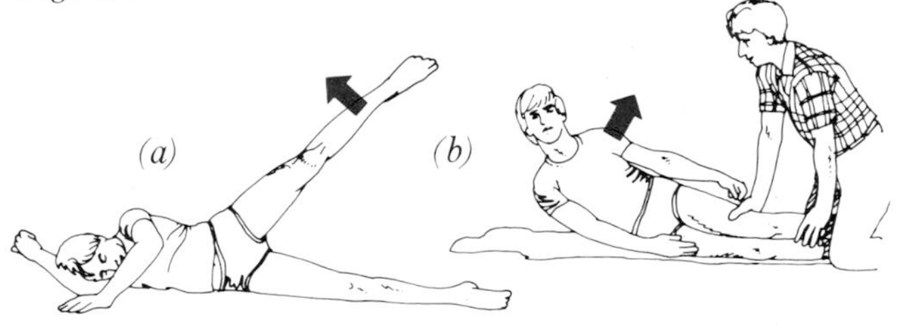

The large buttock muscle (gluteus maximus) is used to swing the leg backwards powerfully, and it can help to keep the knee straight. This is possible because a part of the muscle is attached to the surface of the thigh bone (gluteal tuberosity; straightens the hip) and a part is inserted into a very strong, thick tendon band on the outer side of the thigh (iliotibial tract). This tendon band passes, in turn, in front of the axis of motion of the knee and is inserted into the lateral condyle of the shin bone. The tendon band is felt as a flat 3–4 cm wide tendon on the outside of the thigh directly above the knee joint.

The large buttock muscle can work with greater force if the body is bent forward at the hip. This is because the distance between its origin and insertion becomes greater (see p.18).

Fig. 90.

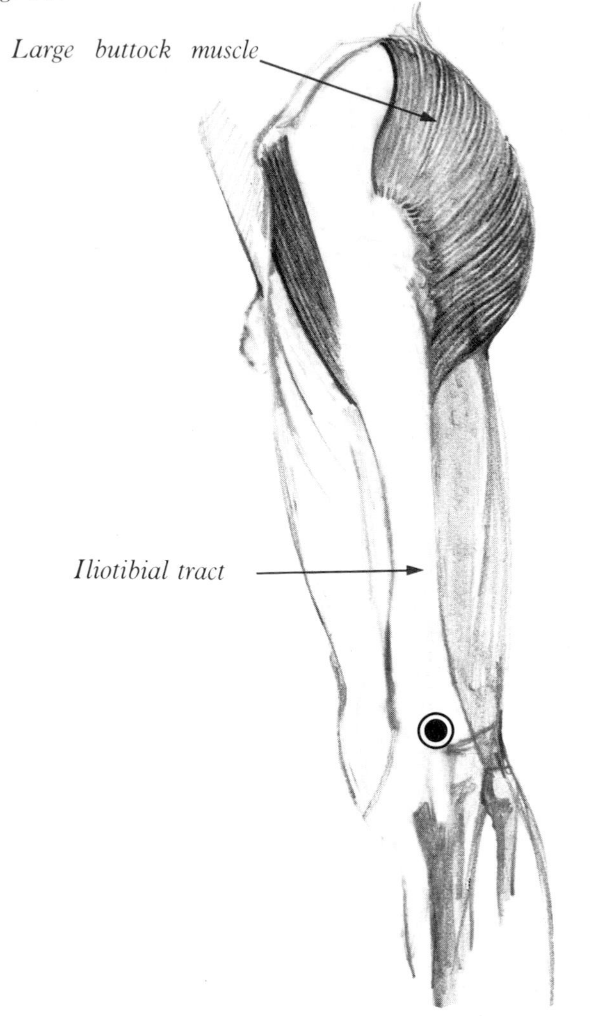

The more force a person needs, the further forward he must lean. The best movements for training the large buttock muscle are those that simultaneously engage the knee and hip extensors.

Fig. 91.

The muscles which take part in swinging the leg backwards (hip extension) are — besides the large buttock muscle — those muscles originating from the ischial tuberosity. These muscles are all inserted into the lower leg which means that they flex the knee joint (see p.53).

2. The groin muscles (adductor muscles) swing the leg towards the midline of the body. They have received their names according to their area of origin, size and appearance.

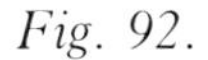

Fig. 92.

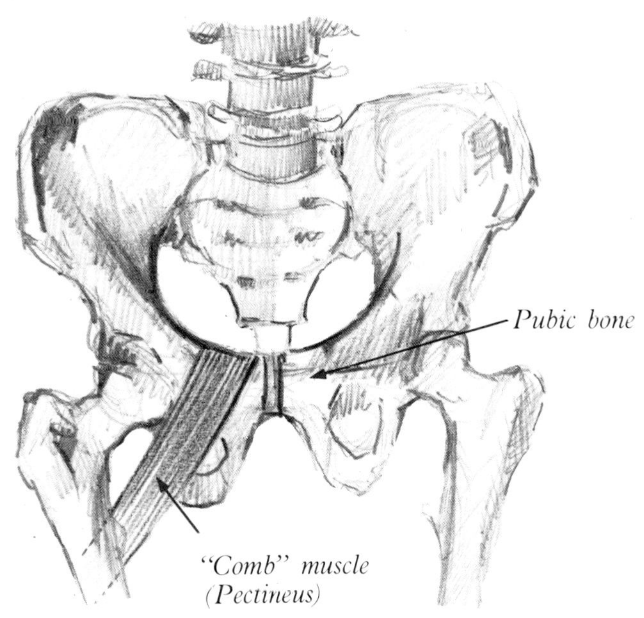

All these muscles originate from the pubic bone (os pubis) and are inserted into the posterior surface of the thigh bone via the roughened ridge extending the length of its shaft (linea aspera). These muscles work powerfully when, in running, the foot leaves the ground and begins to swing forward. During the forward swing, the leg rotates outwards in relation to the hip. This can be accomplished because the adductor muscles are inserted into the posterior surface of the thigh bone. Such overexertion as occurs in the forceful movements of broad side-kicks in football, bringing the free leg forward in skating, tough sprinter training, etc., leads to discomfort in the muscle's area of origin (groin injuries). Groin injuries can be avoided by developing muscular strength and, more importantly, flexibility (p.25). Some strength-developing exercises are shown in Figure 97.

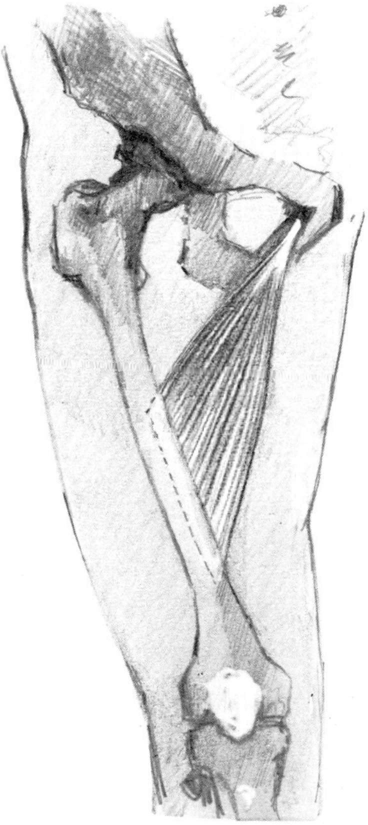

Fig. 93.

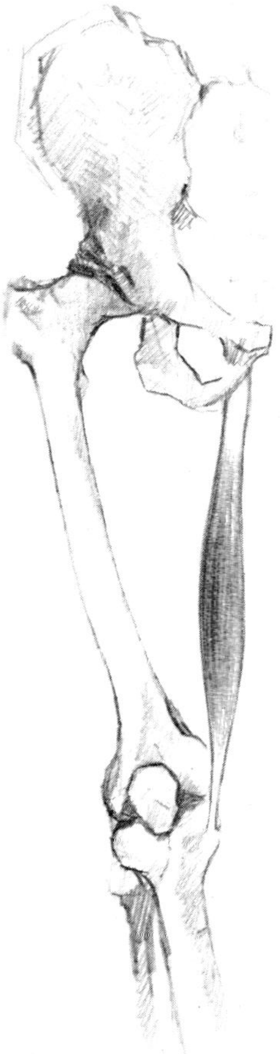

Slender thigh muscle (Gracilis)

Fig. 94.

Long adductor (adductor longus)

Fig. 95.

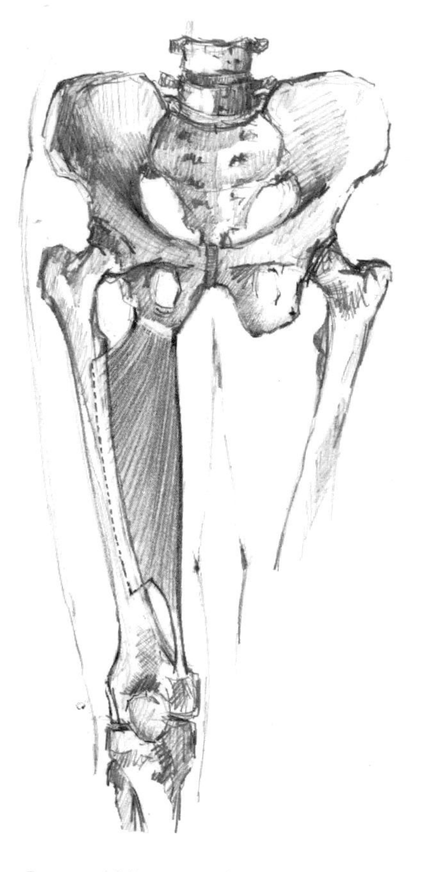

Large adductor (adductor magnus)

Fig. 96.

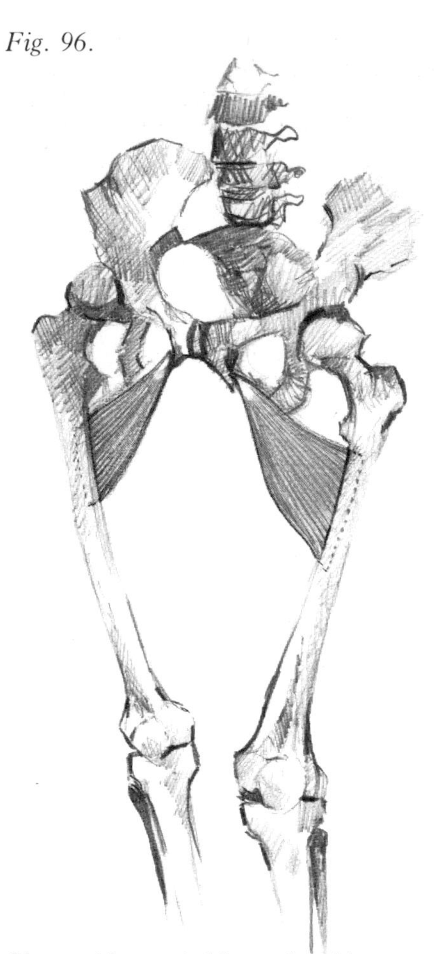

Short adductor (adductor brevis)

Fig. 97 (a).

Fig. 97 (b).

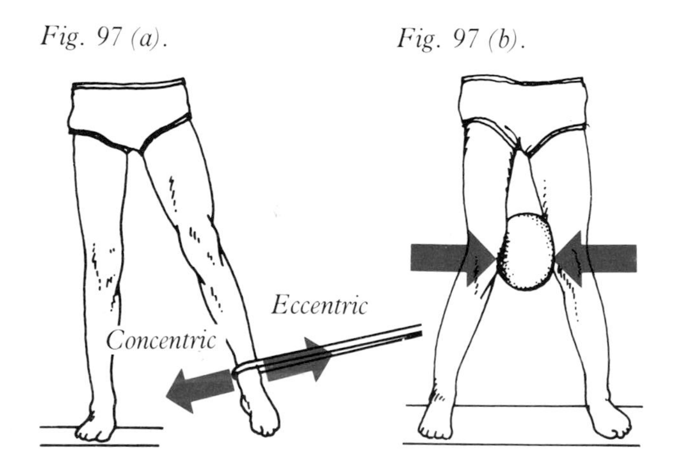

Concentric training when the legs are brought together. Eccentric training when one leg is slowly drawn away from the other.

For static training, compress a ball with both knees.

Fig. 97 (c).

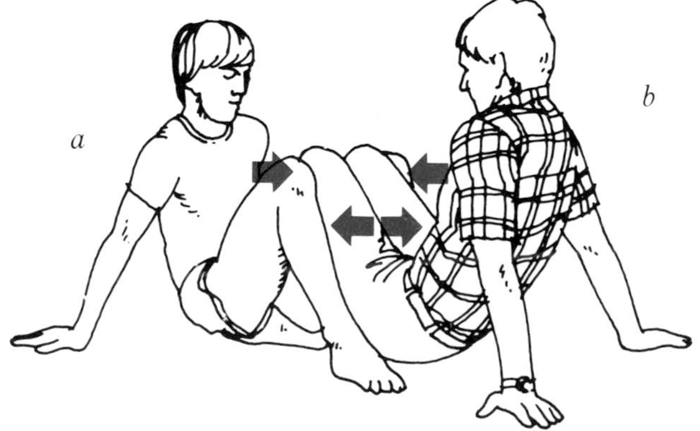

a Trains the adductors, b trains the abductors.

Outward rotation is affected by a number of small muscles which originate from the inner parts of the pelvis [Figure 98(a)]. They pass behind the thigh bone and are inserted into its outer surface at the greater trochanter. They are used a great deal in such an activity as ice-skating.

Fig. 98

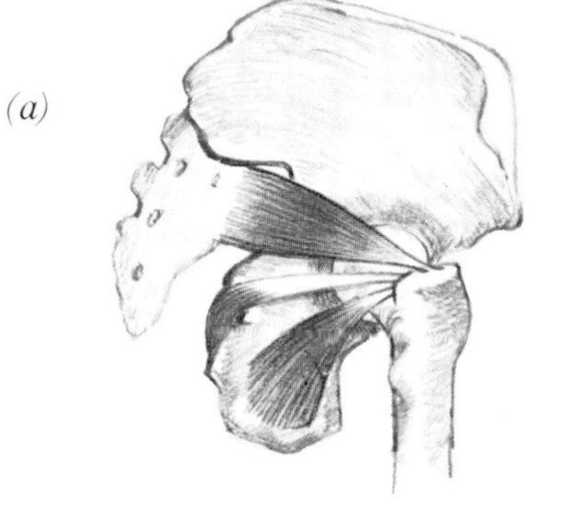

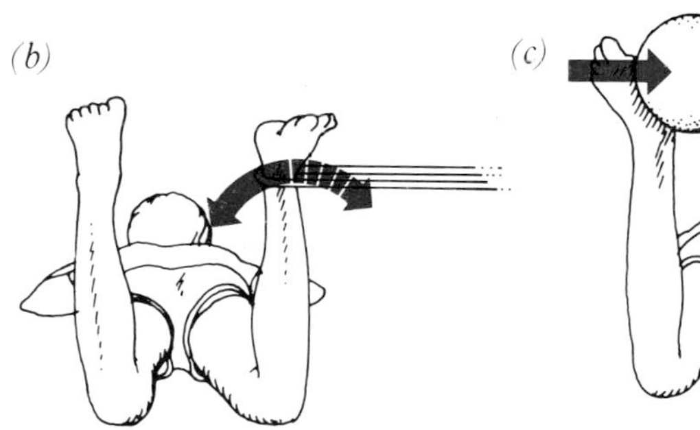

Example of dynamic training of the outward rotators.

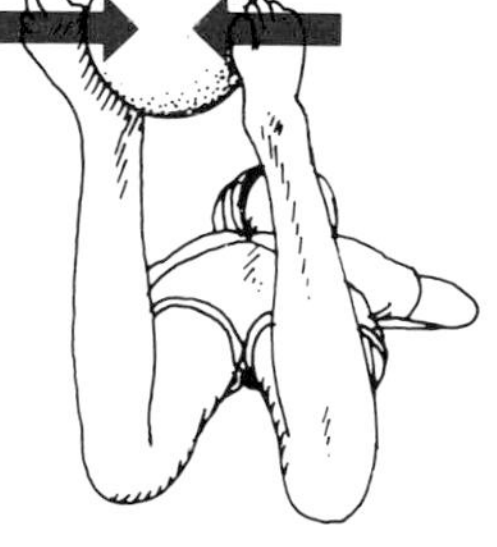

Example of static training of the outward rotators. Lie on your stomach and try to "squash" the ball with your feet.

3. Hip flexors

The haunch muscle (iliacus) and great lumbar muscle (psoas major) are responsible for powerful flexion at the hip joint. They have different points of origin, but a common insertion point. They are often described under the collective name of the iliopsoas.

Fig. 99.

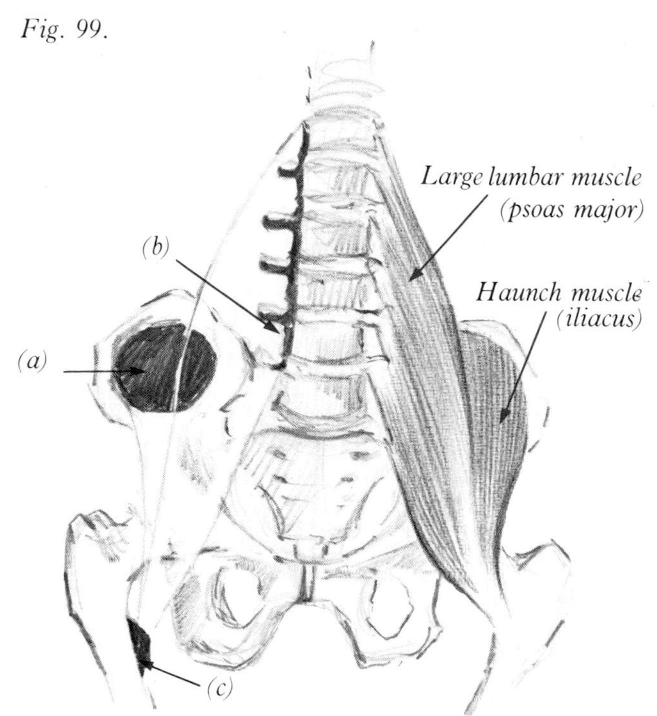

(a) Origin of iliacus (iliac crest).
(b) Origin of psoas major (lumbar part of the vertebral column).
(c) Insertion of both (lesser trochanter).

The following can occur when the iliopsoas contracts.
(a) If the legs are fixed, the trunk will move towards them as, for example, in the last phase of a sit-up [Figure 104 (a) and (b)].

Fig. 100.

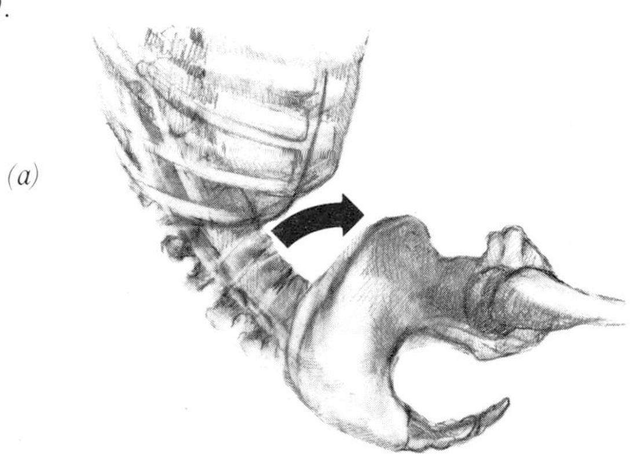

(b) If the trunk is fixed, the legs will move towards it as, for example, when hanging from a bar and trying to bring the knees up towards the chest [Figure 104 (c), (d), (e) and (f)].

Fig. 101.

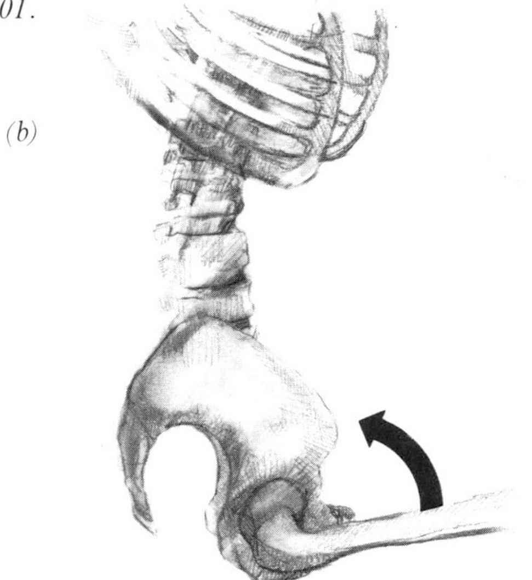

The iliopsoas is incomparably the most powerful hip flexor. It is forcefully engaged in the following examples.

Fig. 102.

It is not necessary to train the iliopsoas for everyday use since it is sufficiently trained in such activities as walking, running, climbing stairs, etc. Athletes must realise that the stress they subject this muscle group to when they train it for strength affects both the insertion and origin (e.g. lumbar vertebrae) to the same extent. Thus, the spine tends to sway forwards which subjects the discs between the vertebrae (p.64) to great stress. This action is counteracted by the abdominal muscles which hold the spine in place. The abdominal muscles are capable of correcting a sway of the lumbar spine by their contraction. Thus, the iliopsoas should not be subjected to greater stress than the abdominal muscles are capable of "parrying". Training ought to be aimed primarily at building up the abdominal muscles (p.71).

Fig. 103.

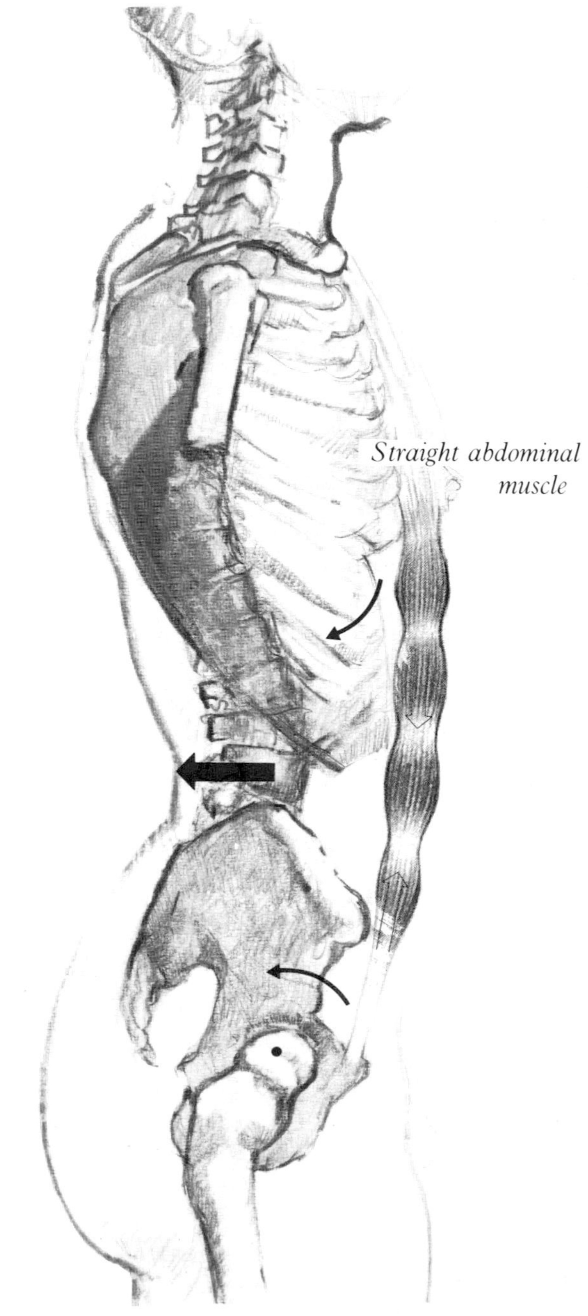

After building up the abdominal muscles, the hip flexors can be specially trained. Here are some examples of exercises which specifically train the hip flexors of athletes whose abdominal muscles are well-trained: (prophylactic training of the abdomen is dealt with on on p.74).

Fig. 104.

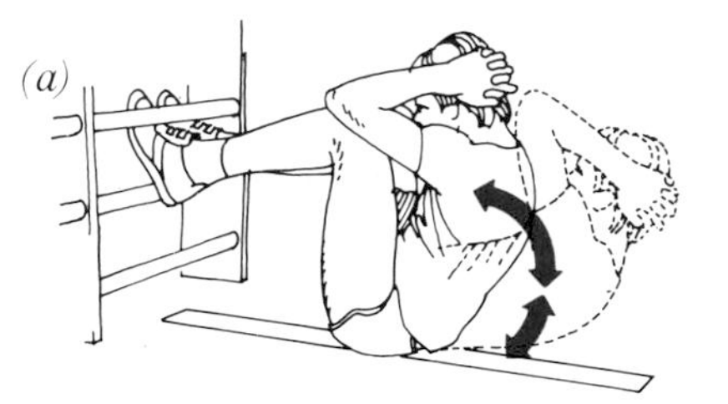

(a) Sit-ups with bent knees. Weights (2 kg, 5 kg, . . .) may be held against the chest to increase the work load. The first three-quarters of the sit-up is pure abdominal muscle training (p.74). The rest of the movement concentrates on the hip joint and thus trains the iliopsoas for strength.

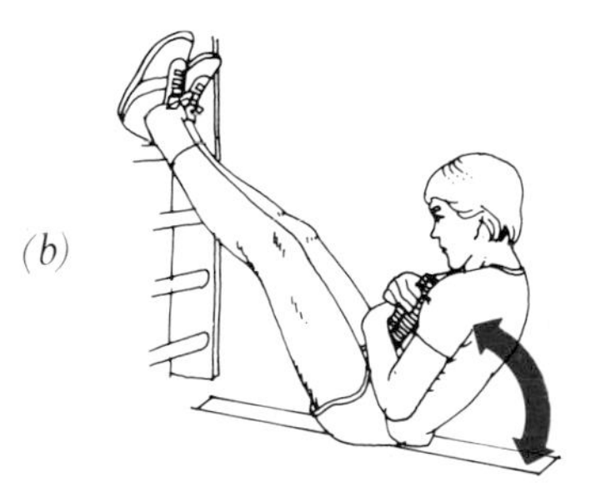

(b) If the legs are kept straight, then the exertion will be even greater due to the resistance offered by the muscles at the back of the thigh.

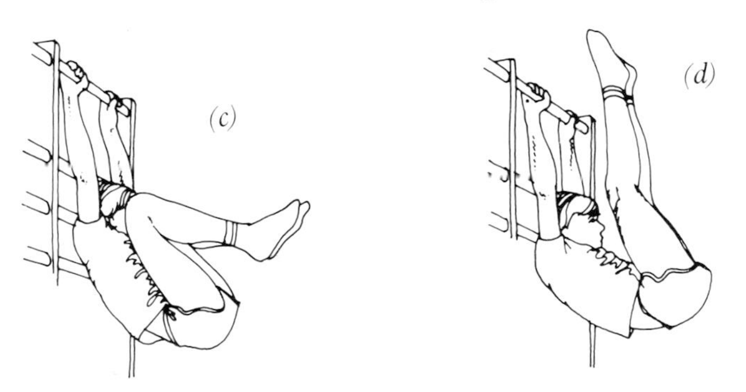

(c) Lie on a slanting plane or hang from a bar and lift your knees towards your hands, or

(d) (for greater stress) lift your legs as high as possible, keeping them straight. For eccentric training, resist the movement on the way down.

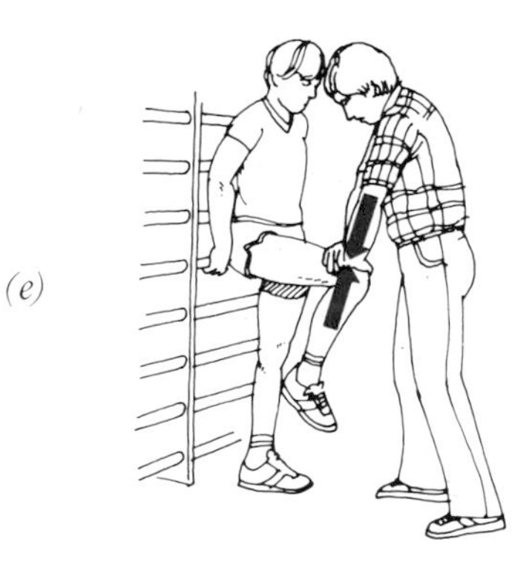

(e) Stand by a wall and have a friend provide external resistance to your thigh as you try to lift your knee.

(f) Tie a band of rubber around your foot and pull your knee up in quick movements.

(g) The hip muscles can be trained statically in the following way. Lie on your back with a 90° angle at your hips and have a friend take a hold of the lower part of your shin. By making large movements in different directions, he should try to make you change the angle between your legs and upper body.

In order to stretch the iliopsoas, the distance between its origin and insertion must be lengthened as much as possible while the muscles are relaxed. One way to stretch them is to kneel in the manner shown in Figure 105, i.e. the stress is placed on the leg in front and the legs are separated as much as possible. If the iliopsoas is short and strong, but the muscles at the back of the thigh are weak, then the body tends to tip forward at the pelvis. This leads to mild back-aches and a posture resembling that of a person with a "beer belly".

Fig. 105.

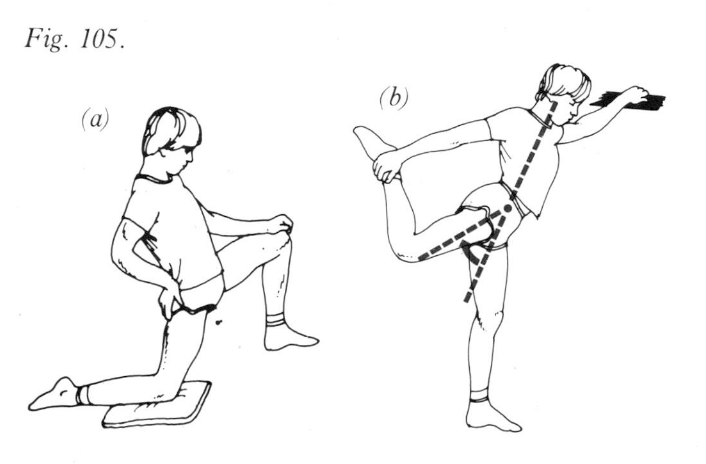

Another stretching exercise that has an effect on the iliopsoas is the following. Stand on one leg. With your hand pull your free leg back behind an imagined line passing vertically through the centre of your body. Let the backward movement take place at the hip and not in the lower back. Keep a wide angle in the knee joint, otherwise the movement will be resisted by another muscle, namely the straight thigh muscle (rectus femoris, p.49).

Fig. 106.

A third stretching exercise is shown in Figure 106. A friend lifts your leg upwards and backwards at the same time as he holds your pelvis still (thereby blocking any movement of the lumbar spine). This exercise should only be carried out by people who have received their instructions from a medically qualified trainer.

Fig. 107.

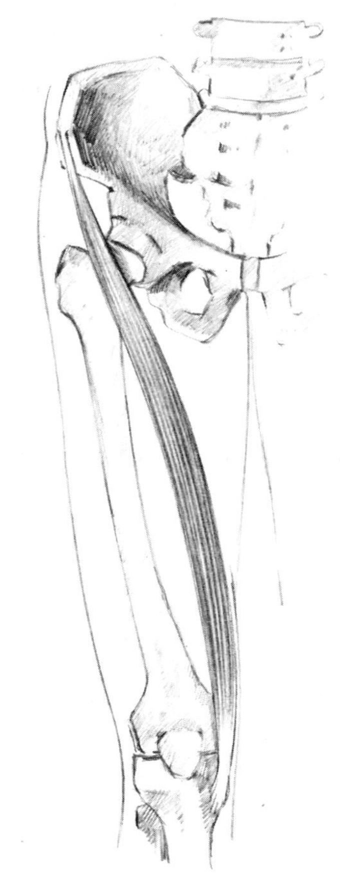

The tailor's muscle (Sartorius) is the longest muscle in the body (Figure 107). It extends from the anterior superior iliac spine (spina iliaca anterior superior) and passes down the thigh in a slightly S-shaped curve to the inner side of the knee where it is attached to the internal condyle of the tibia (Condylus medialis tibiae).

This muscle has so many different functions that it is difficult to place it in any particular muscle group. The tailor's muscle received its name from the fact that its functions allow a person to sit on a table with crossed legs as tailors once did. Thus, this muscle bends, abducts and rotates the hip outwards, and bends the knee and rotates it inwards.

C. The Knee Joint

The knee joint is a very complicated joint which needs detailed study. The movements produced at the knee joint are bending (flexing) and straightening (extending), and inward and outward rotation of the lower leg. The last-mentioned movements can only occur when the knee is bent. The more the knee is bent, the easier it is to rotate the lower leg and the foot. We can think of the knee as flexing either by:

Fig. 108.

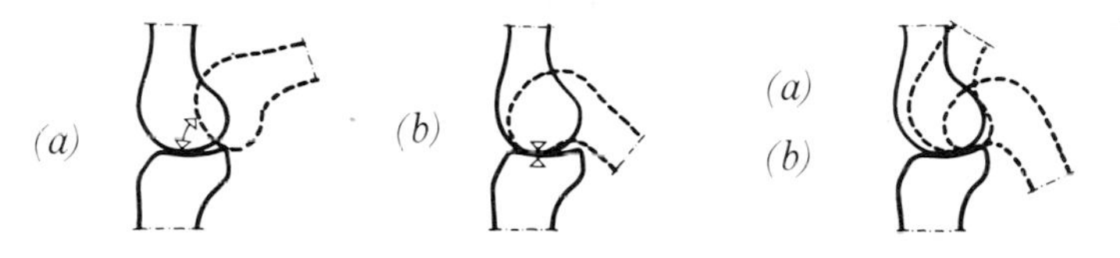

(a) the thigh bone (femur) rolling back on to the shin bone (tibia) or by (b) the thigh bone gliding on the same spot on the shin bone.

In reality, both types of movement occur. Movement (a) takes place until the anterior of the two crossed ligaments (anterior cruciate ligament) is completely stretched, after which movement (b) is brought into action.

Thus, the anterior cruciate ligament prevents movements when the lower leg is moved forward in relation to the thigh. A common injury in football is rupture of the anterior cruciate ligament when a player receives a powerful kick on the back of his lower leg. The posterior cruciate ligament is injured when the lower leg is pressed backwards or when the knee is severely overstretched.

Fig. 109.

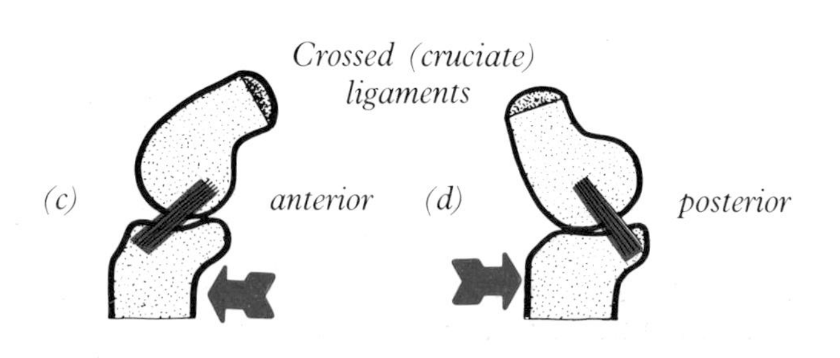

The function of the two side (or collateral) ligaments is to prevent sidewards bending of the knee. They are taut when the knee is stretched (e), and slack when the knee is bent (f). This means, for example, that the lower leg can be rotated outwards until the ligaments are again taut. The lower leg usually cannot rotate as much inwards as it can outwards because the cruciate ligaments in the joint "twist" around each other during inward rotation and thereby block the movement (see Figure 20).

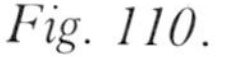

Fig. 110.

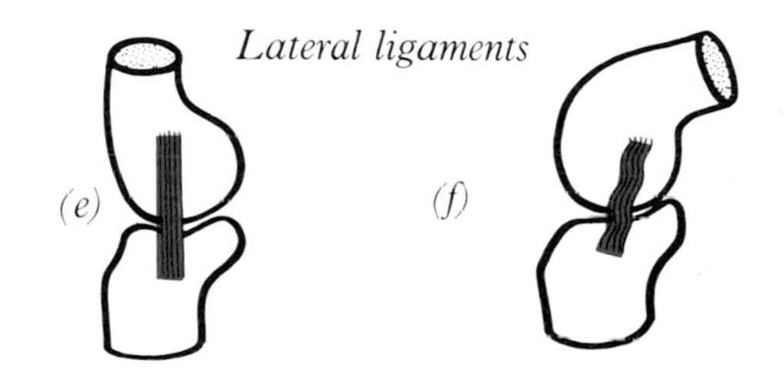

The lower end of the thigh bone is ellipsoid and the upper extremity of the shin bone is flat. Therefore, there would be very little contact between their surfaces if the cartilage were not so thick and the so-called menisci were not shaped to receive the end of the thigh bone. The underside of the menisci is plain like the surface of the shin bone. Consequently, the stress to which the knee is subjected can be distributed over a relatively large area.

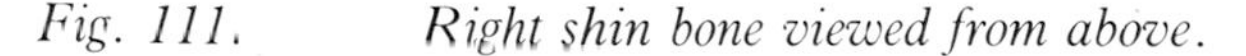

Fig. 111. *Right shin bone viewed from above.*

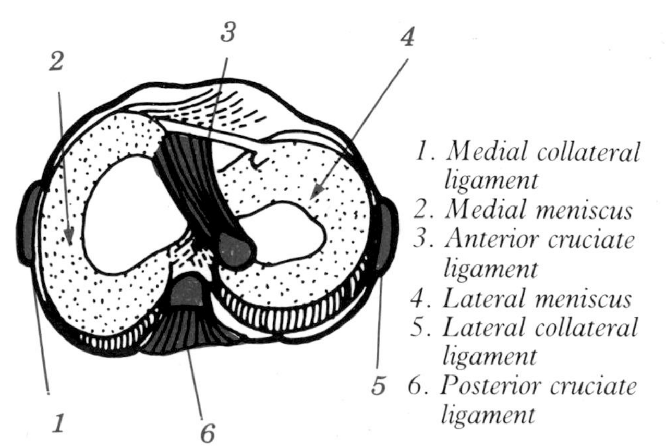

In flexion and extension of the knee joint, the menisci glide to suit the form of the condyles of the thigh bone. Because the medial meniscus fuses with the medial collateral ligament, it is very easily injured by being subjected to excessive stress while in "strange" positions.

Injury to the menisci due to rotational strain on a bent weight-bearing knee is very common. If there is a sudden pitching in the knee joint during outward rotation of the lower leg, the medial ligament stretches and can thereby tear the meniscus, which is locked between the thigh and shin bones. Because of this fact, movements of the following type should be avoided: (a) crow hopping or walking like a duck and (b) lying in the hurdle position to train flexibility.

Fig. 112.

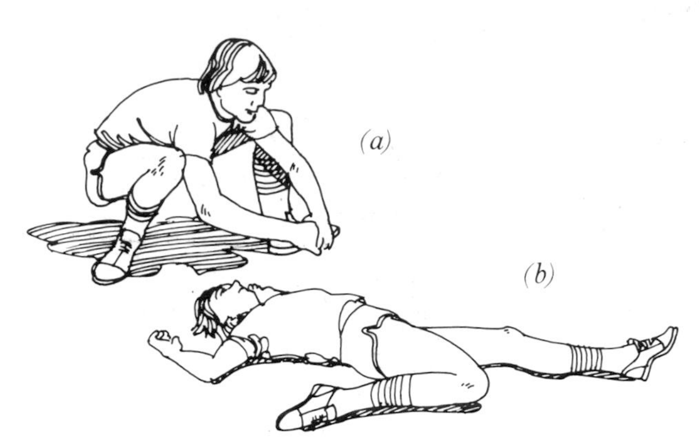

D. Muscles of the Knee Joint

Knee extensors

The straight thigh muscle (rectus femoris) originates from the pelvis and bends (flexes) the hip joint. It is inserted into the kneecap and can — via the powerful tendon which extends from the kneecap to the shin bone — straighten the knee.

Fig. 114.

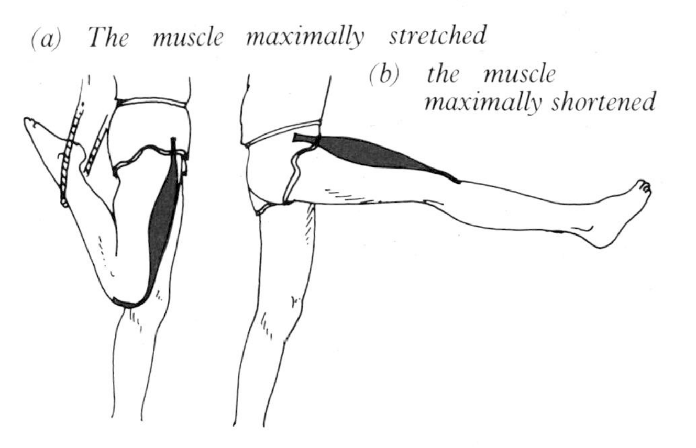

(a) is by far the best position for stretching the straight thigh muscle in accordance with the PNF-method.

Fig. 113.

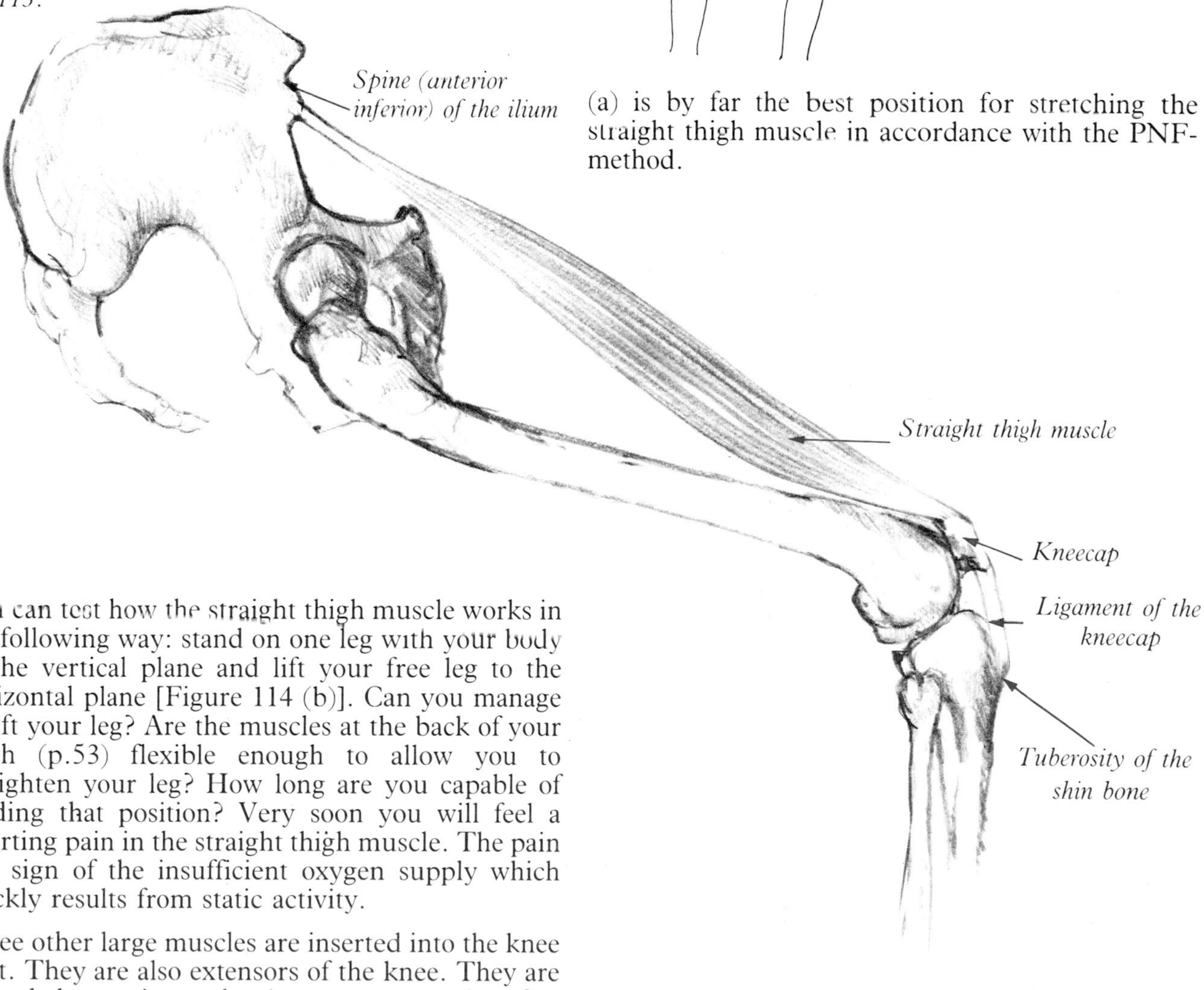

Spine (anterior inferior) of the ilium

Straight thigh muscle

Kneecap

Ligament of the kneecap

Tuberosity of the shin bone

You can test how the straight thigh muscle works in the following way: stand on one leg with your body in the vertical plane and lift your free leg to the horizontal plane [Figure 114 (b)]. Can you manage to lift your leg? Are the muscles at the back of your thigh (p.53) flexible enough to allow you to straighten your leg? How long are you capable of holding that position? Very soon you will feel a smarting pain in the straight thigh muscle. The pain is a sign of the insufficient oxygen supply which quickly results from static activity.

Three other large muscles are inserted into the knee joint. They are also extensors of the knee. They are termed the vasti muscles (vastus = extensive, far-reaching). The four diagrams of Figure 115 show the extensors in the right leg viewed from the front.

Fig. 115.

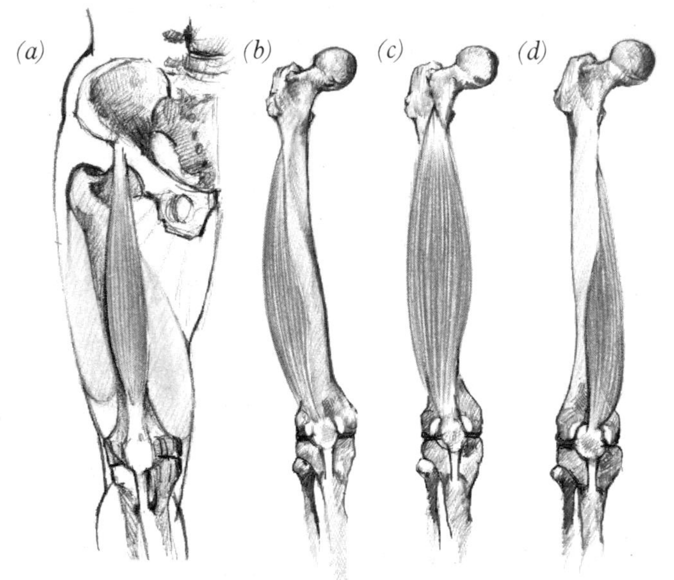

Origin:

(a) The straight thigh muscle (rectus femoris).
Origin: Anterior inferior iliac spine.
(b) The external vast thigh muscle (vastus lateralis).
Origin: Thigh bone ridge (linea aspera).
(c) The central vast thigh muscle (vastus intermedius).
Origin: Anterior thigh bone shaft.
(d) The internal vast thigh muscle (vastus medialis).
Origin: Thigh bone ridge (linea aspera).

Insertion:
Some of the fibres of the internal thigh muscle are inserted into the kneecap. The remaining muscles attach via a tendon into the kneecap. This is, in turn, attached to the tubercle of the shin bone (tibial tuberosity) via the kneecap ligament (patellar ligament) (Figure 113).

Function:
Straightens the knee when working concentrically. Impedes flexing of the knee joint when working eccentrically.

The collective name for the three vasti muscles and the straight thigh muscle is the four-headed thigh muscle (quadriceps femoris). Their action is to straighten the knee and to stabilise and guide the kneecap so that it correctly glides in the depression formed by the condyles of the thigh bone (p.52).

The exercises shown in Figure 116 use the quadriceps femoris.

Fig. 116.

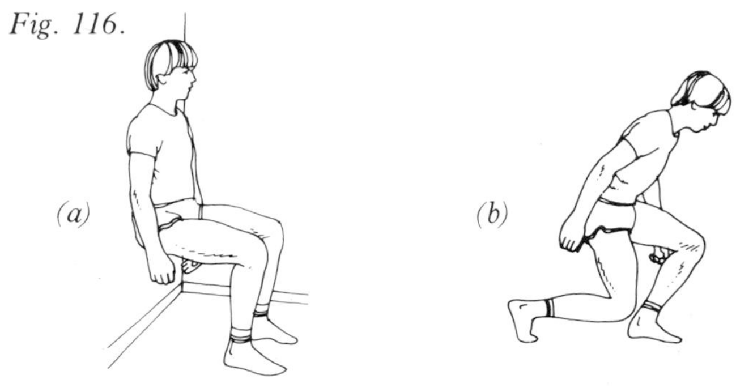

(a) Sit with your back against a wall with a 90° angle between your hip and knee joints. The muscle works statically.

(b) Stand with your weight on your left foot and place your right knee directly behind the tendon of your left heel. You will notice a distinct difference if you stand completely still in this position (static stress), or if you bounce lightly up and down (short periods of relaxation allow the blood to be supplied to the muscle).

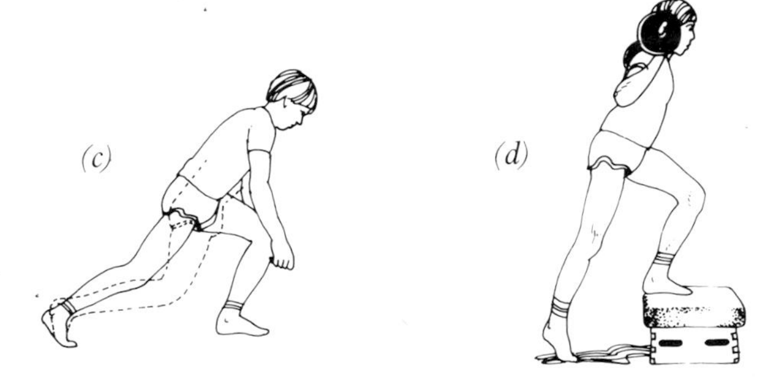

(c) Start walking with long strides followed by elastic stretches accompanying each stride. Your bodyweight should be placed far out over your left foot. The halting movement trains the muscle eccentrically.

(d) Step up and down from a bench or part of a vaulting horse and — if your back is sufficiently strong — you could even carry a load on your shoulders. This exercise provides dynamic (mostly concentric) training.

Fig. 116.

(e) Jump up (concentric work) and down (eccentric work) on the spot. You must halt the movement before bending your knees too much (not less than a 90° angle at the knee joint, see Figures 117 and 118).

The work load in exercises (a)–(e) is greatest when the knee is flexed. The deeper you descend, the greater the work load will be. The muscles are subjected to large but harmless stress. However, the cartilage at the back of your kneecap will be subjected to harmful stress if you bend your knees too much. An explanation follows. (See p.18 for diagrams and descriptions of how the origin and insertion of muscles are affected by muscle forces.)

Examples of stress produced in the knee joint

(a) When standing on one leg (with slightly bent knee and centre of gravity 5 cm behind the axis of motion of the knee joint), the knee extensors (quadriceps femoris) must contract with enough force to keep the body from collapsing and sitting down. According to the seesaw principle, the force of the muscle must equal that of gravity if the muscle's lever arm is also 5 cm.

(b) If the knee is flexed at a lesser angle than in (a), the distance between the vertical line [Figure 117 (b)] and the axis of motion of the knee will become 15 cm; although the muscle's lever arm will remain unchanged at 5 cm. In example (b) the muscle force must be 3 x 600 N because its lever arm is three times less than that of the gravitational force.

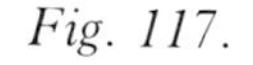

Fig. 117.

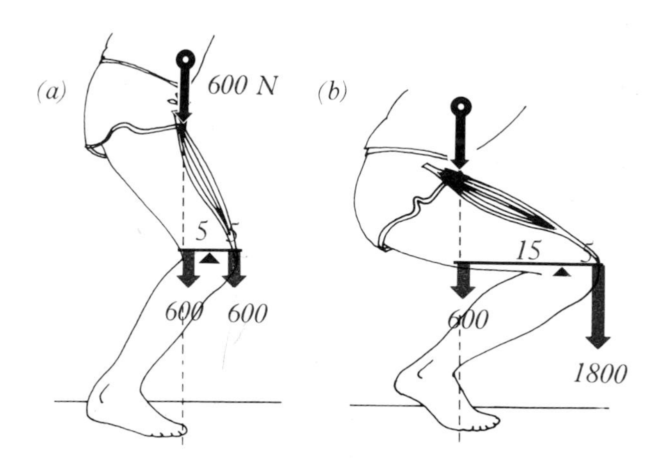

With extreme bending of the knee a muscle force of 4–5 times the force of gravity (bodyweight) may be required. The force that pulls the kneecap over the thigh bone (femur) can thus amount to 3000 N. Consequently, the kneecap is subjected to forces that together press it against the thigh bone with a force that, at small angles, is almost double — i.e. 5000 N according to Figure 118. If exercises of this type are repeated frequently, the athlete runs a great risk of tearing the cartilage. The rule of thumb: "never" bend too deeply at the knees while they are subjected to stress (in the form of extra weights, speed, short stopping distance, etc.), is really worth following!

The calculations above show that the stress to which a straight knee is subjected is almost zero. They also show that the deeper you bend, the greater the stress will become.

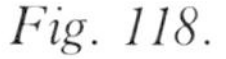

Fig. 118.

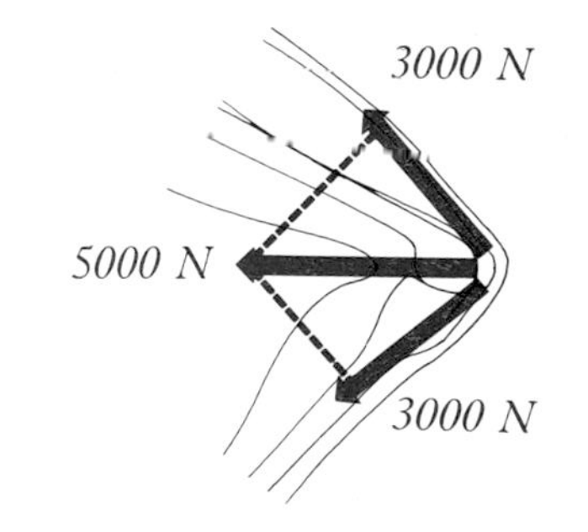

With the help of an EMG (electromyography = the recording of muscle activity), it has been shown that the internal thigh muscle (vastus medialis) is most active during the final extension of the leg. In order to train this part of the quadriceps, it is necessary to construct exercises that put the greatest stress on the leg when the knee is fully extended. None of the exercises described on p.50 are of this type.

In the following exercises (Figure 119) the extended knee is subjected to intense work loads (force to overcome).

Fig. 119.

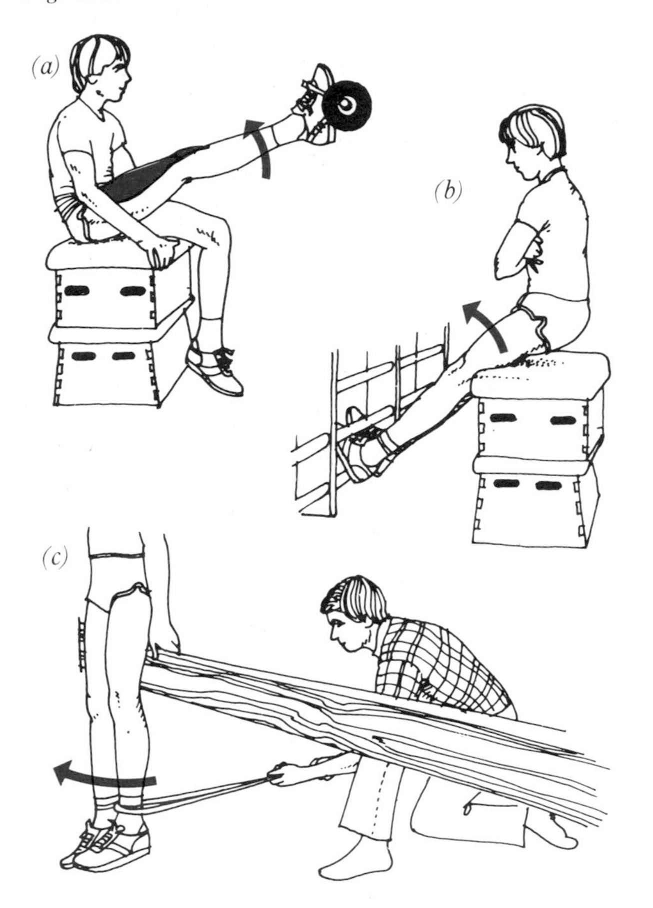

In all three exercises the work load is greatest when the knee is fully extended (the knee remains extended for static stress). These exercises particularly strengthen the internal vast thigh muscle, a muscle which is particularly important for the stability of the knee. The thigh bone and the shin bone form a certain angle with each other. The kneecap is acted upon primarily along the direction of the thigh (rectus femoris and the medial and lateral vasti), but the depression in which the kneecap glides, lies on a vertical line (see Figure 120).

The muscle whose job it is to guide the kneecap correctly, so that it is not drawn outwards and "scrape" against the external or lateral condyle of the femur, is the internal vast thigh muscle (vastus medialis). Its direction of pull is shown by the arrows in Figure 120. This muscle is weakened extremely quickly when the knee is injured. For this reason, a special training programme in which the extended knee is subjected to stress should be given top priority after the knee has been in a cast for a time. The diagrams below (Figure 121) illustrate some relatively easy exercises that are aimed at building up strength after an accident. Their goal is to place the greatest work load on the extended knee.

Fig. 120.

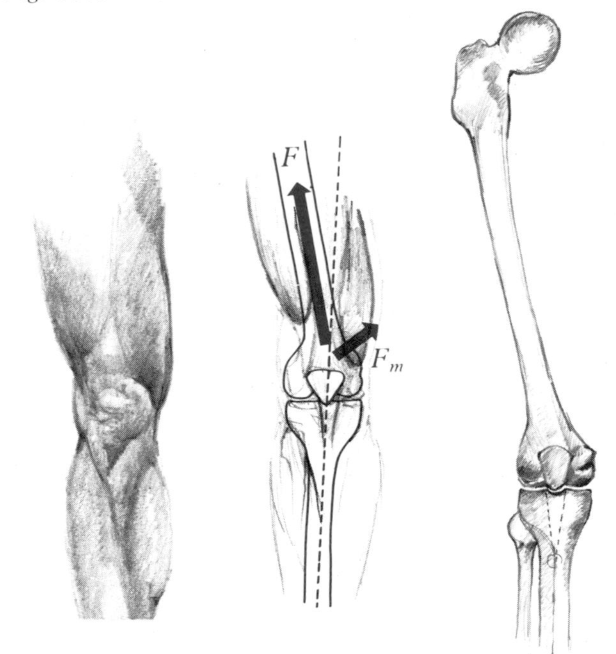

Fig. 121.

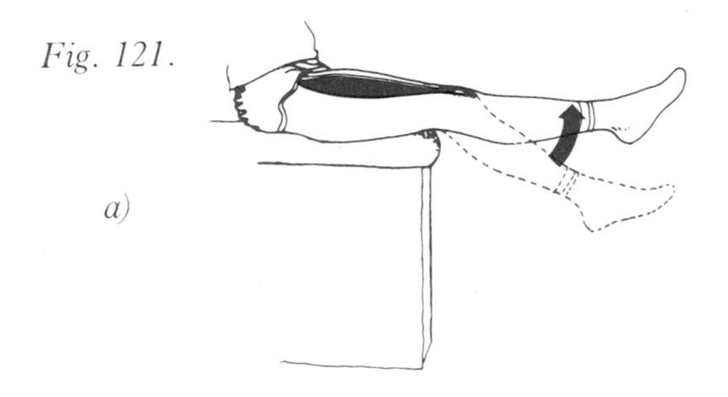

(a) The four-headed thigh muscle must overcome the weight of the lower leg. Its work load can be increased by placing a sandbag on the foot or by wearing a weighted shoe.

Fig. 121.

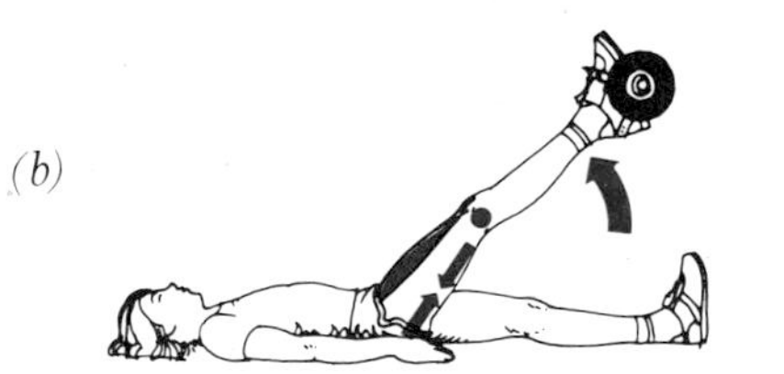

(b) If the leg is lifted while straight, the weight of the lower leg still subjects the four-headed thigh muscle (quadriceps femoris) to stress. This exercise is more demanding than (a) because the muscles at the back of the thigh stretch and strive to flex the knee joint. Thus, the extensor has to work harder than in example (a).

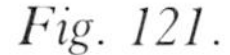

Fig. 121.

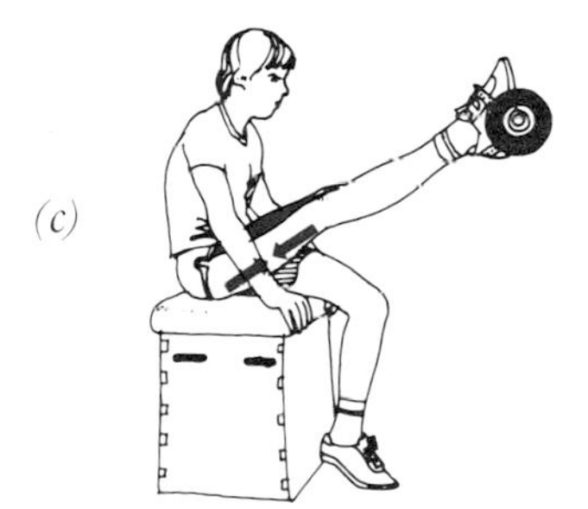

(c) When a body sits, the pelvis tips forwards, which does not happen if it lies as shown in example (b). This causes the muscles at the back of the thigh to stretch even further. Thus, exercise (c) is heavier than exercise (b). Furthermore, it also subjects the flexors of the hip to stress.

A well-trained elite high jumper, for example, could aim at being able to sit and lift a straight leg 45° above the horizontal plane with a load of 15 kg on his foot.

Knee flexors

When judging the value of different exercises designed for the knee extensors, one must take account of their antagonist, i.e. the muscles situated at the back of the thigh. The collective name for these muscles is the hamstrings. They originate from the lowermost part of the hip bone (ischium) and run towards the knee. There are three in all.

Fig. 122

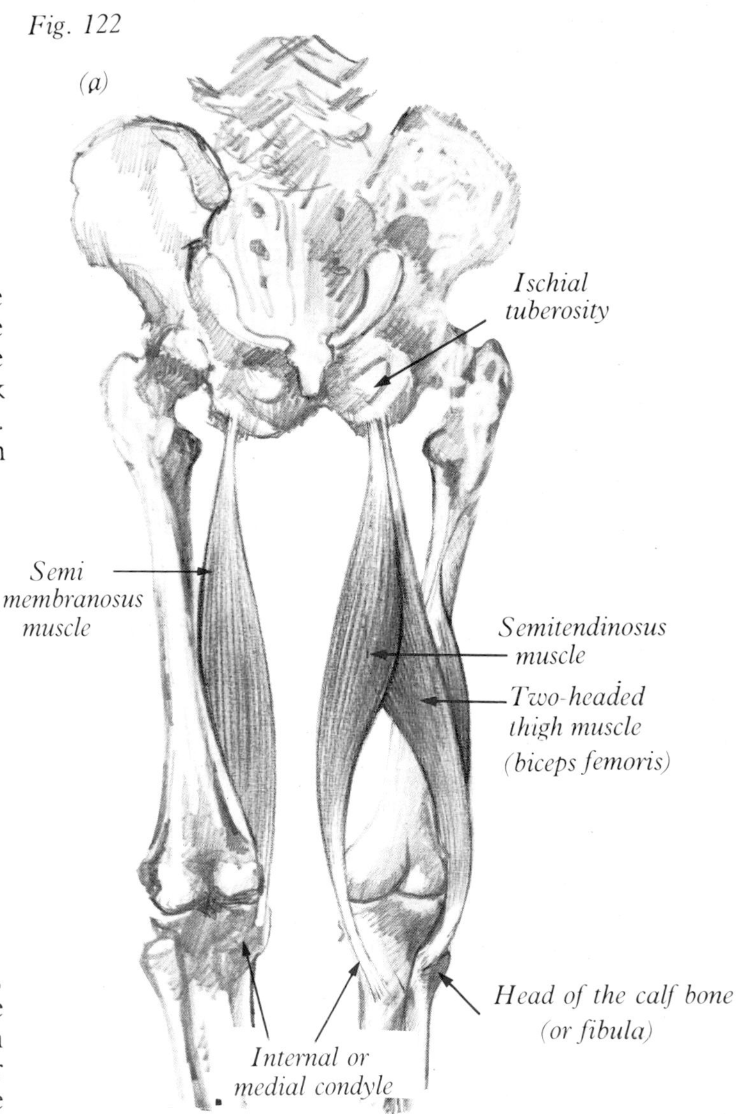

The pelvis and femur seen from behind

All three originate from the ischial tuberosity. The two-headed thigh muscle (biceps femoris) is inserted into the head of the calf bone (or fibula). It can rotate the lower leg so that the foot points outwards. The semitendinosus and semi-membranosus muscles are inserted into the internal or medial condyle of the shin bone (tibia); therefore they can rotate the lower leg inwards.

This group of muscles is called the hamstrings because their tendons can easily be felt at the back of the thigh. The distance between the origin and insertion of these hip extensors and knee flexors vary greatly depending upon the angle of the hip and knee joints.

Fig. 122 (b).

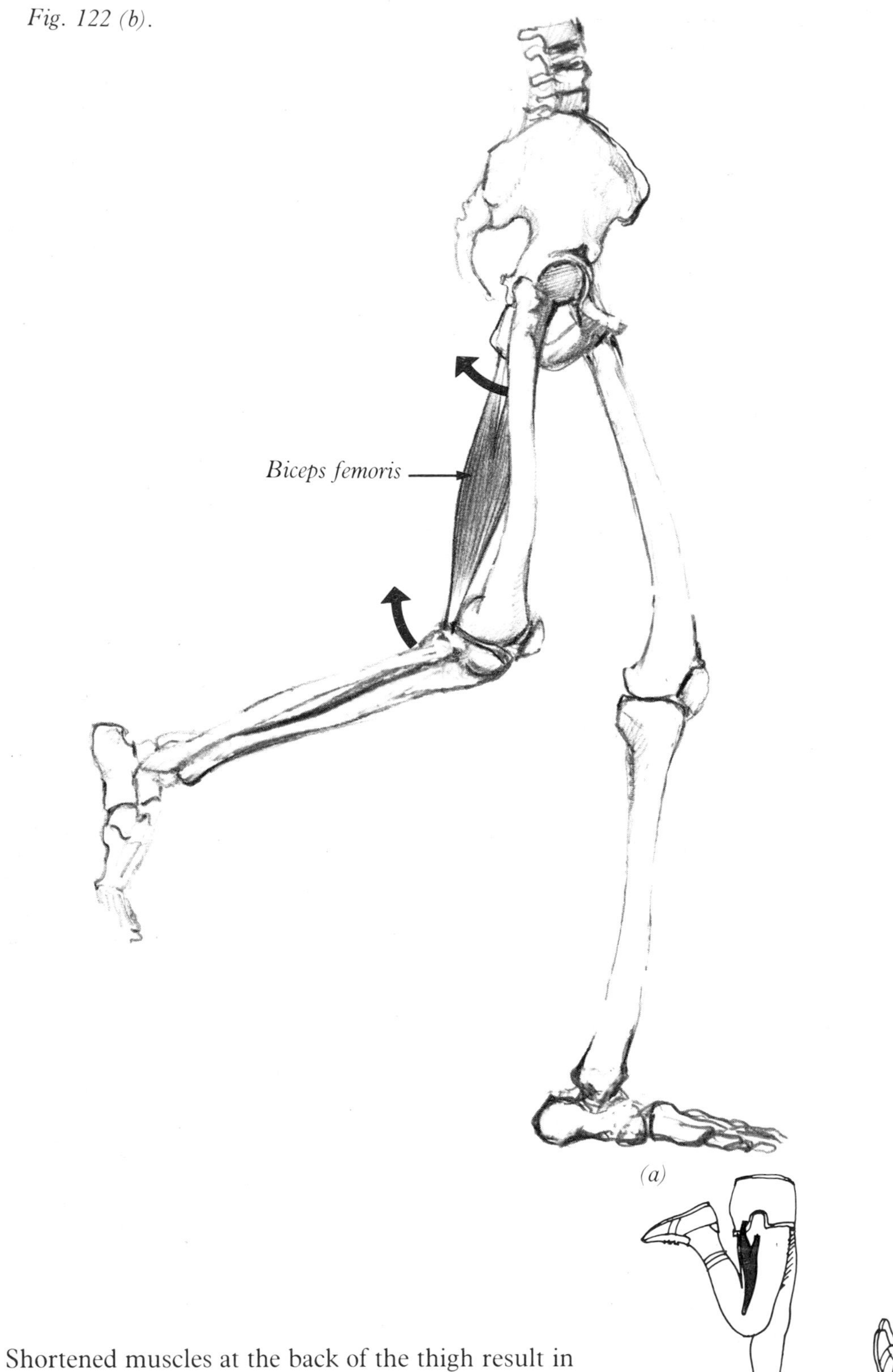

(a)

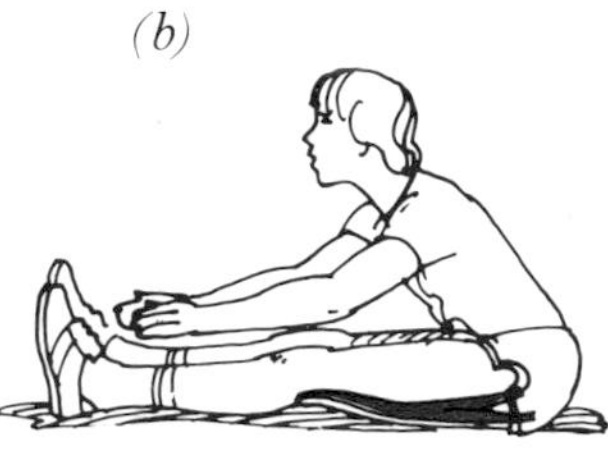

(b)

Shortened muscles at the back of the thigh result in a stationary hip. When a person is unable to tip his pelvis forward, he tries to compensate for this by bending forward at his lumbar spine. Back trouble often results from the hamstrings being too short.

Fig. 123.

(a) Maximally contracted (b) Maximally stretched

Flexibility exercises for the knee flexors

The following diagrams (Figure 124) demonstrate stretching exercises for the knee flexors using the PNF-method (p.25).

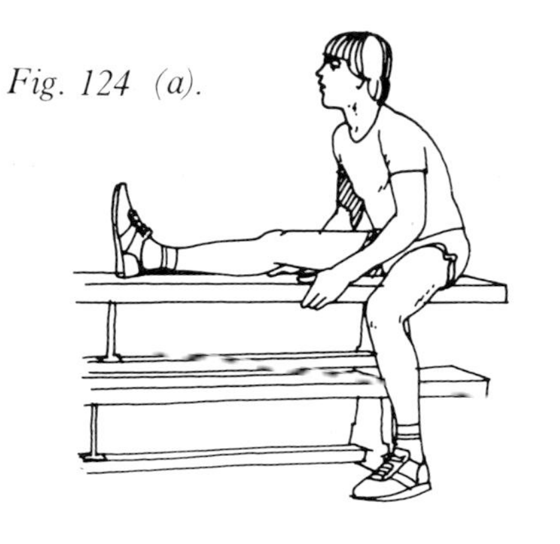

Fig. 124 (a).

(a) Assume this position [Figure 124 (a)] without bending forward at your lumbar spine, and keep your back straight. Tip your pelvis as far forward as your thigh muscles will allow.

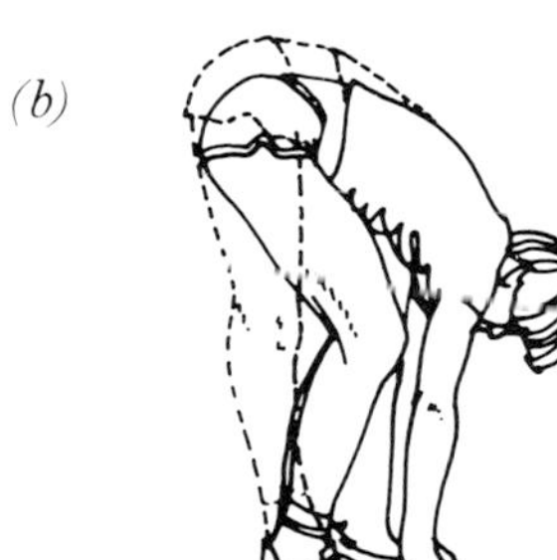

(b)

(b) Stand with your knees bent and your hands touching the ground. Then stretch your knees until you feel a comfortable tension at the back of your thigh. By supporting your hands on the ground you are taking the load off your back. If you straighten one leg at a time, the lifted leg will further unload your back. This exercise is best suited for those who already have relatively long hamstrings.

Fig. 124.

(c)

(c) Lie on your back and pull one leg — held straight — towards your chest.

Strength-developing exercises for the knee flexors

The exercises below are examples of different strength-developing exercises for the back of the thigh.

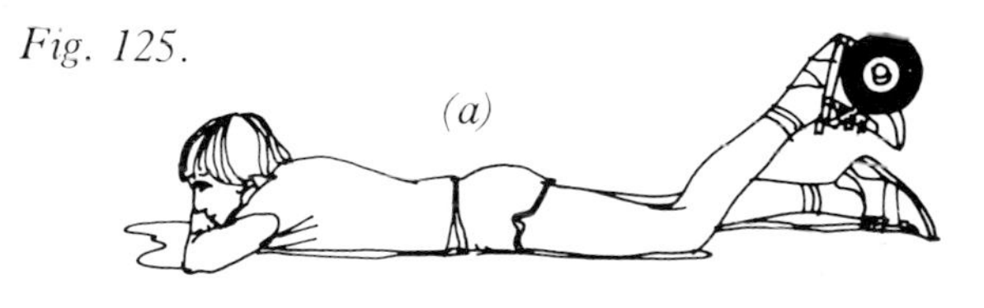

Fig. 125.

(a)

Figure 125 (a). A friend (or weighted shoe or rubber hose) should provide just enough resistance to allow you to bend your knee with even movements backwards (eccentric work for the muscles) and forwards (concentric work). The resistance should not be so great that you are forced to bend at the hip in order to perform the movements.

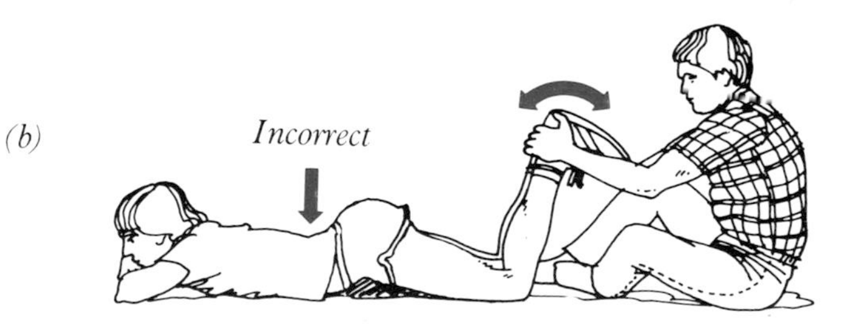

(b)

(b) By bending at the hip using the iliopsoas (i.e. tip your pelvis forward) the origin of the hamstrings (the ischial tuberosity) will move further away from your lower leg. Thus, you stretch your flexors and thereby increase their strength (p.18). This means that you subject your lumbar spine to stress in an unfavourable position, and this could lead to pain. Thus, you ought to press your hip down and not strain it more than is necessary to perform the movement with your hips touching the ground.

Fig. 125.

Keep your back flat

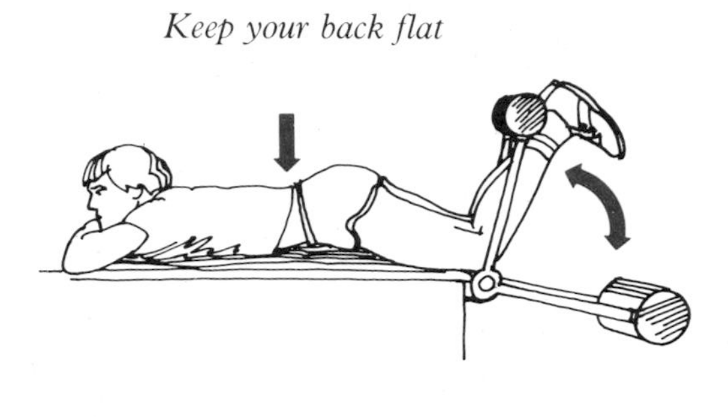

(c) The same strength-developing apparatus that was used to train the knee extensors can be used to train the knee flexors.

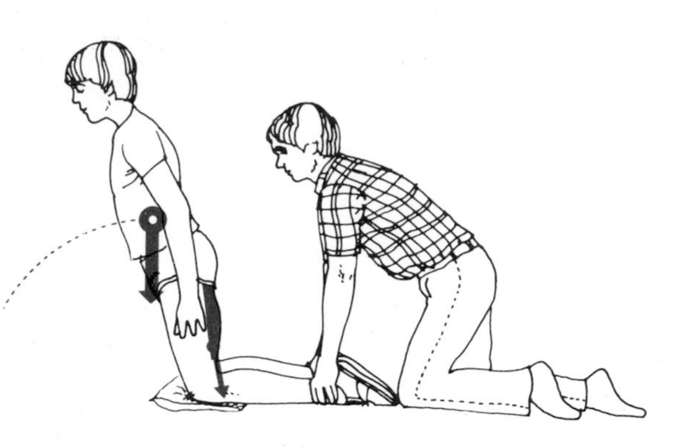

(d) The following exercise is demanding. Kneel on a soft surface with your lower legs fixed. Then let your upper body fall slowly forward and thereafter pull it back: Keep your hips straight. It is advisable to begin with small movements. This exercise subjects the muscles to great stress and can easily lead to a tendency for cramps if one is untrained. Here, it is the body's own weight that subjects the muscles to stress. The further the body's centre of gravity is placed in front of the knee joint, the greater is the moment of force the knee flexors must cancel. When the trunk is lowered, the muscles act eccentrically; when it is raised to the vertical position they are acting concentrically.

In order to train the muscles at the back of the thigh for coordination and speed you can run with short quick strides; sprint and try to kick your buttocks; or cycle quickly (in first gear).

E. Lower Leg and Foot

A muscle group which is very important for jumping and running is the calf muscle (triceps surae) [Figure 126 (a) and (b)]. This muscle has three parts: (a) the "twin" calf muscle (gastrocnemius) with its two heads of origin (one from each of the condyles of the thigh bone) and (b) the "flounder" muscle (soleus) (a flat muscle originating from the back of the lower leg). Together the three parts form the heel tendon (tend Achilles) which is inserted into the heel bone (calcaneum).

The gastrocnemius bends (or flexes) the knee and the ankle so that the body can be raised on its toes (plantarflexion). The soleus muscle acts only on the ankle joint.

Fig. 126.
(a) Twin calf muscle (gastrocnemius)
(b) Flounder muscle (soleus)

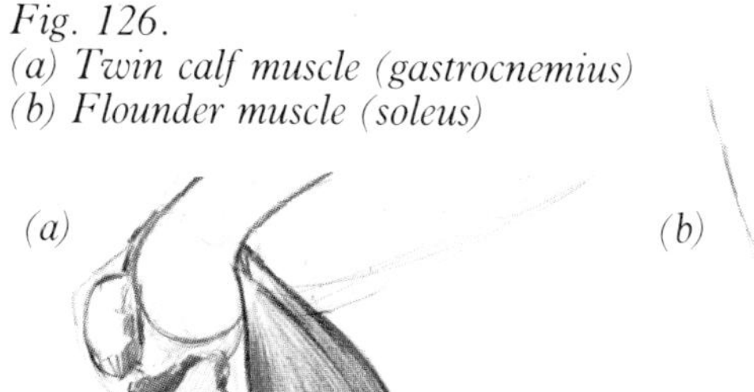

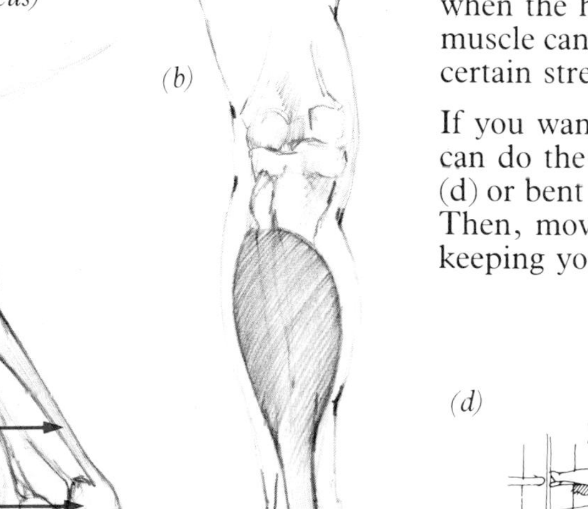

(c) Place the balls of your feet against something that will raise them 5 cm above your heels. Stand on your toes, and then return to your original position. This exercise can be done quickly many times (dynamic endurance) or with a heavy weight five or six times (maximal strength training) (p.29). When pressing upwards the muscles are trained concentrically and when descending they are trained eccentrically. For a very short period — when the heel is in contact with the ground -- the muscle can relax in its extended position. This has a certain stretching effect.

If you want to stretch only your calf muscles, you can do the following exercise. Stand against a wall (d) or bent forward (e) with support for your hands. Then, move one leg as far as possible behind you keeping your heel on the ground.

The muscle can be activated with the PNF-method by pressing the ball of your foot against the ground for a few seconds. Another good way is to plant the ball of your foot against a tree and pull yourself carefully forward with your arms so that the calf muscles are stretched (f). If your knees are straight, then you will also be stretching your gastrocnemius, (d), (e) and (f). If your knees are bent, the soleus is stretched (g). The latter is often neglected. When it is, the calf muscle still feels stiff and sore.

Fig. 126.

The foot

When the calf muscle is too short, the foot tends to assume a position with downward-pointing toes (Figure 127). The muscles which hold the foot up (situated at the front of the lower leg between the shin and calf bones, p.60), are thus forced to work constantly with raised tension (tone) in order to hold the foot in its normal position. Such tension may lie behind the pain which is experienced in the lower leg when an athlete trains too much at one session, or runs on hard surfaces, etc.

Fig. 127.

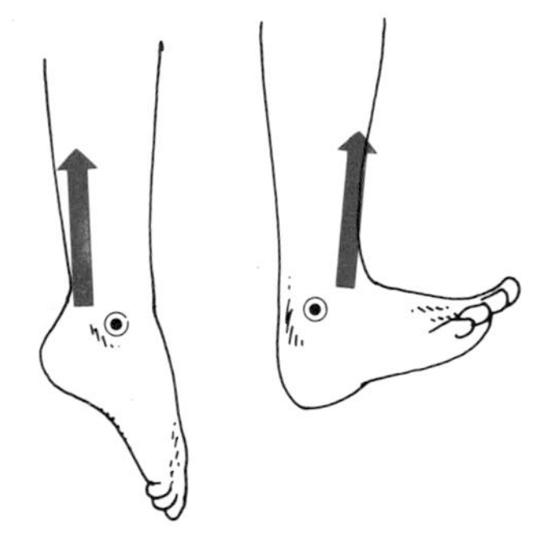

When we take a closer look at how the heel tendon is attached to the heel bone, we can easily understand how mechanically correctly the muscles of the body are arranged. If the heel tendon were attached as shown in Figure 128 (a), below, then its lever arm (i.e. ability to turn the ankle) would have less and less affect the more the body is raised on its toes (d_1 would be reduced to d_2).

Fig. 128.

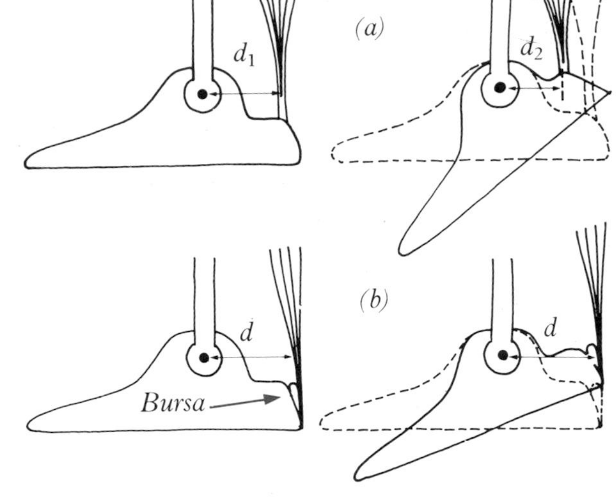

However, the heel tendon is attached to the heel bone near its base, which means that the lever arm *(d)* remains about the same length regardless of whether the whole foot is firmly planted on the ground or whether the body is standing on its toes.

Fig. 129.

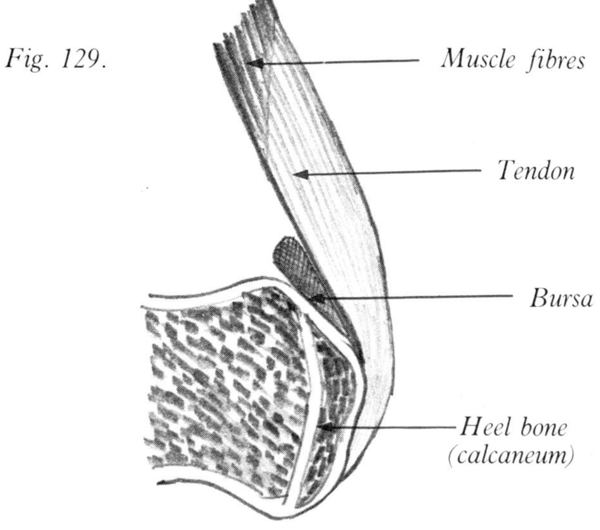

The tendon is protected from scraping against the heel bone by a bursa which is situated between the tendon and the bone.

Movements of the foot

The foot can move around two axes. The movements around axis 1 are called flexion and extension. Movements around axis 2 are called supination and pronation.

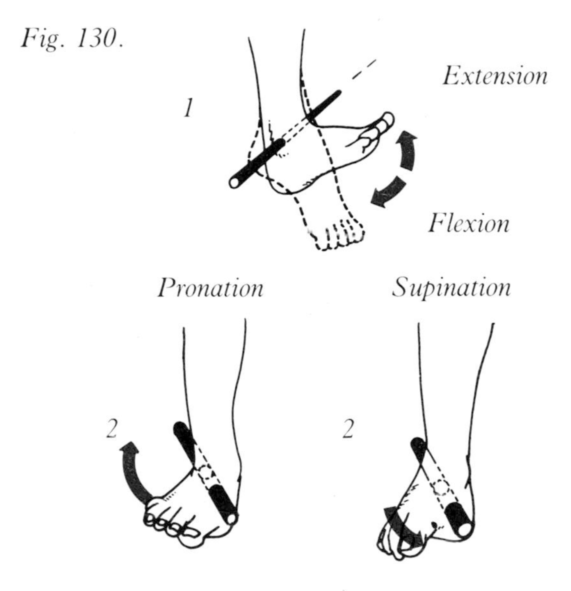

Fig. 130.

Skeleton of the foot

The skeleton of the foot is divided into ankle bones (tarsus), bones of the foot (metatarsus) and the bones of the toes (phalanges).

Extension and flexion take place between the ankle bone itself (talus) and the lower leg bones (i.e. tibia and fibula). This joint is called the ankle joint.

Supination and pronation take place between the talus and the "boat-shaped" tarsal (the navicular) bone and the heel bone (the calcaneus). This joint is known as the subtalar joint.

Supination and pronation take place simultaneously between several articulating surfaces which together form the subtalar joint.

Movements of the ankle and subtalar joints are independent of each other. The movements of the foot at these joints are controlled, as usual, by the muscles. If the muscles are unable to prevent movements that are too great or too sudden, the joint is still protected by the ligaments of the foot.

The ankle ligaments arise from the prominent lower ends (the malleoli) of the shin and calf bones (the tibia and fibula) and spread out down towards the articulating ankle bones. Thus, the internal or medial ligament of the ankle arises from the medial malleolus of the shin bone (tibia) and is inserted into the calcaneus, talus and navicular bones. It is called the deltoid ligament.

Fig. 131 (a).

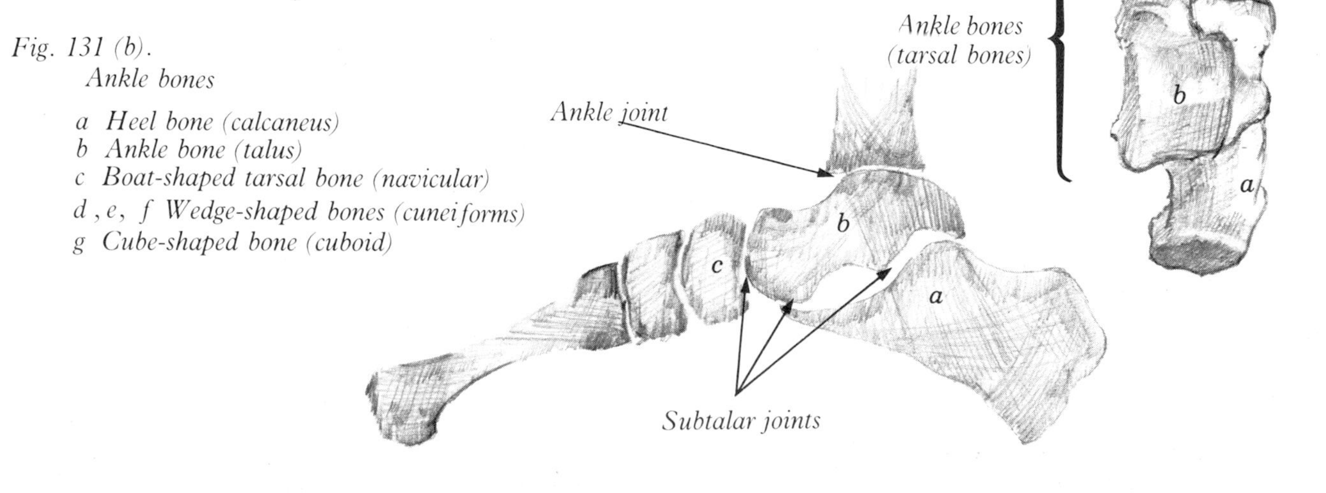

Fig. 131 (b).
Ankle bones

- *a Heel bone (calcaneus)*
- *b Ankle bone (talus)*
- *c Boat-shaped tarsal bone (navicular)*
- *d, e, f Wedge-shaped bones (cuneiforms)*
- *g Cube-shaped bone (cuboid)*

There are three separate ligaments making up the external or lateral ligament. They all arise from the external or lateral malleolus of the calf bone (fibula). One of them extends forward and is inserted into the talus, one passes downward to be attached to the calcaneus, and the third one extends backward towards the back part of the talus.

(1, 2, 3, in Figure 133) are found at the front of the leg between the shin and calf bones. Their tendons can easily be felt on the top of the foot at the base of the shin bone. Pronation is brought about by the two muscles whose tendons can be felt beneath the external or lateral (calf bone) malleolus (4, 5). Supination is produced mainly by the three muscles whose tendons pass behind and beneath the internal or medial (shin bone) malleolus (7, 8, 9). Figure 133 (b) shows their locations in relation to the different axes of motion.

Fig. 132.

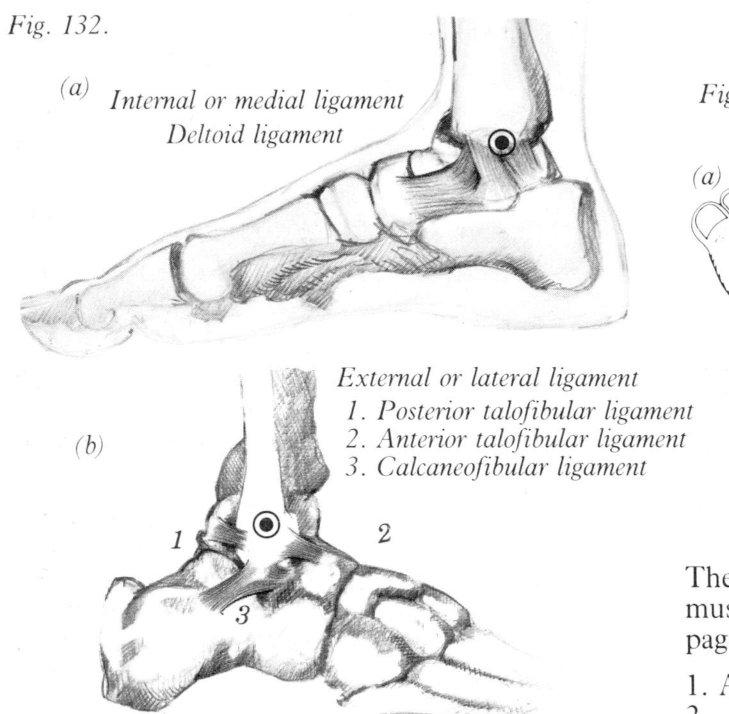

Fig. 133.

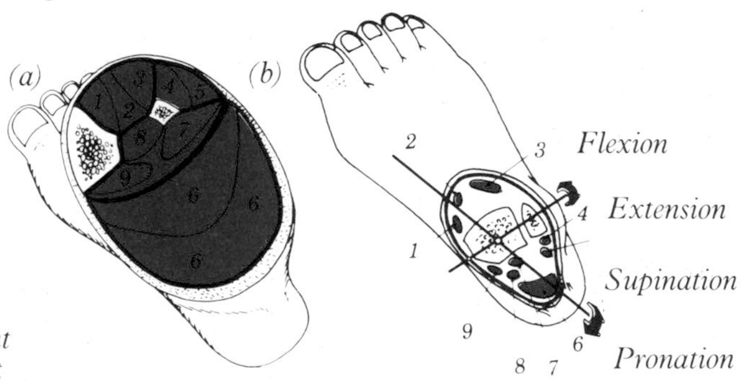

The origin of the deltoid ligament corresponds to the axis of motion ◉ of the ankle joint. Thus it is always taut. The origin of the external lateral ligament is situated below the axis of motion. Hence, its posterior part is taut when the foot points upwards and the anterior part is taut when the ankle is stretched. When injured, the ligament either suffers a partial rupture or is completely torn. In many cases the ligament remains intact but the malleolus is broken off.

The numbers in the diagrams refer to the following muscles. (They are also dealt with on the facing page.)

1. Anterior shin bone muscle (tibialis anterior).
2. Long great toe extensor (extensor hallucis longus).
3. Long toe extensor (extensor digitorum longus).
4. Long calf muscle (peroneus longus).
5. Short calf muscle (peroneus brevis).
6. Three-headed calf muscle (triceps surae).
7. Long flexor of the great toe (flexor hallucis longus).
8. Long toe flexor (flexor digitorum longus).
9. Posterior shin bone muscle (tibialis posterior).

Muscles of the foot

The most important flexors (see p.57) are the calf muscles (triceps surae). They are, however, assisted by other muscles whose tendons can be felt and seen behind the malleoli. The most important extensors

When describing the structure and function of the foot it is useful to review the concept of *"arches"*. There are three types.

I Internal (medial) longitudinal arch (movement arch).
II External (lateral) longitudinal arch (support arch).
III Transverse arch (or anterior arch).

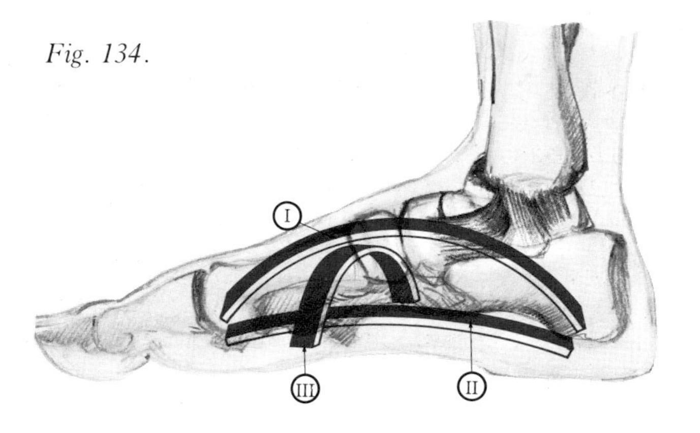

Fig. 134.

Running subjects the external arch to stress. Notice how a running shoe wears down at the outer border of the heel and under the great toe. The other arches — which are supported by *ligaments*, the *wedge-shaped construction of the bones* and the *muscles* — give the foot its elasticity. Departures from the normal appearance can either lead to an arch that is too high (not very common) or too low (flatfoot). The latter may be due to the ligament being stretched which, in turn, could be the result of subjecting the foot to severe stress (running on hard surfaces without arch support) or imbalance of muscle strength between the different muscles. The structures which affect the internal arch are listed below. The structures marked with a + have a tendency to raise the internal arch and those marked with a – tend to lower it.

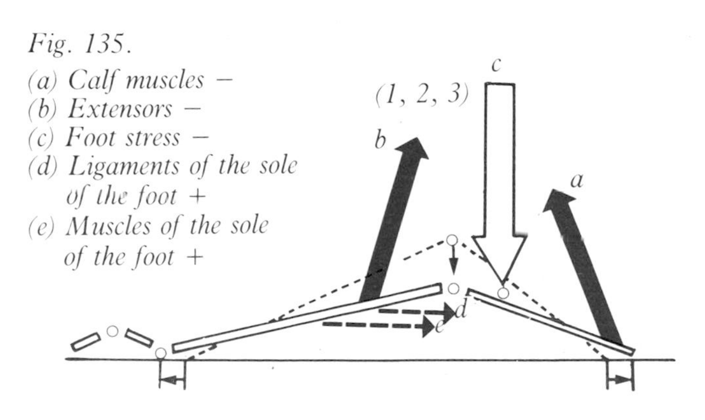

Fig. 135.
(a) Calf muscles –
(b) Extensors –
(c) Foot stress –
(d) Ligaments of the sole of the foot +
(e) Muscles of the sole of the foot +

The most common foot injuries are those sustained by the external or lateral ligament of the ankle joint. These injuries are caused by lopsided gait. The majority of injuries can be avoided by wearing appropriate shoes. Shoes that are worn down on one side only (on the outside of the heel) are usually the culprit. Another common cause of foot injury is a surface that provides so much friction that the shoes stick at the slightest contact. The muscles which simultaneously rotate the foot outwards and pronate (i.e. lift the outer border so that the shoe does not become firmly fixed on the ground too soon) are the long and short calf muscles (peroneus longus and brevis, 4, 5). It is very important that these muscles are strong and can react quickly. The calf muscles and their antagonists (7, 8) can be trained with "toe push-ups".

If you stand (a) on your toes with the sole of your foot pointing inwards as shown below (Figure 136). and push up balancing on the outside of your foot, you will be engaging 7, 8 and 9. If you stand (b) on your toes with the sole pointing outwards, 4 and 5 will be engaged.

Fig. 136.

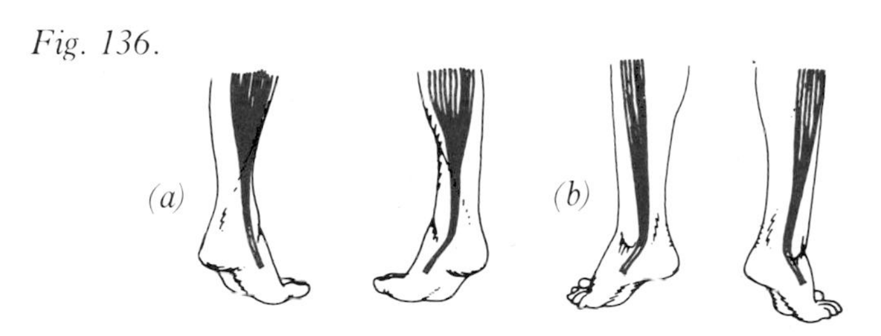

Coordination and speed can be trained by standing and balancing on an unstable object.

You can both control and train your ability to finely adjust your foot position in the following way: stand on one leg with a bent knee keeping your arms by your sides and eyes shut. It is always considerably more difficult to balance on one leg if that leg has recently been injured as it takes time to retrain fine motor ability. The risk of new sprains is obvious if rehabilitation is not hastened.

Fig. 137.

Periostitis

A problem facing athletes sooner or later is inflammation of the membrane covering the bones, i.e. periostitis. The muscles responsible for lifting the foot (1, 2, 3) are situated at the front of the leg between the shin and calf bones. They are enclosed in a fascia (tough connective tissue, Figure 133). The inner part between the shin and calf bones (interosseous membrane) serves as a direct point of origin for some parts of the muscles. The rest of the muscles grow directly out of the bone walls and pass through the bone membrane which has intimate contact with the fascia.

The bone membrane (periosteum) may be detached from the bone due to the force of the muscle contraction or high tension in the muscle fascia which is inserted into the bone through the bone membrane. Small haemorrhages and inflammations appear in the microscopic cavities which form between the detached membrane and the bone.

As mentioned earlier, athletes who train by running a great deal often sustain periostitis. The healing process will be delayed if they continue to run, as by doing so the worn tissues are not given time to heal. Pain can also be caused by changing the running surface or suddenly increasing the amount of training. The muscles — which are confined within the muscle fascia — swell (fill with blood) more than the fascia will permit. The tension that then builds up leads to a pain resembling periostitis pain. Such increased pressure in a muscle compartment is referred to as "closed compartment syndrome".

A successful surgical method for treating severe cases of increased pressure is to cleave the muscle fascia by making a longitudinal slit between the shin and calf bones on the fore side of the leg. In this way the tension of the anterior muscle compartment will be reduced.

Running on hard surfaces can lead to a complaint similar to periostitis. The sudden and excessive stresses to which the leg is subjected at each stride can, at worst, give rise to microscopic small splits in the outer lamella of the bone. This type of complaint heals with greater difficulty than the more common periostitis.

The wearing of unsuitable shoes when running on hard surfaces causes the internal arch to fall at each stride. The anterior shin bone muscle (1) which is attached to the arch (Figure 138) will fall and thus pull on its origin with a great deal of force. This is especially the case when the whole foot is planted on the ground instead of the heel striking first. The shoe should be built up at the arch to give it support.

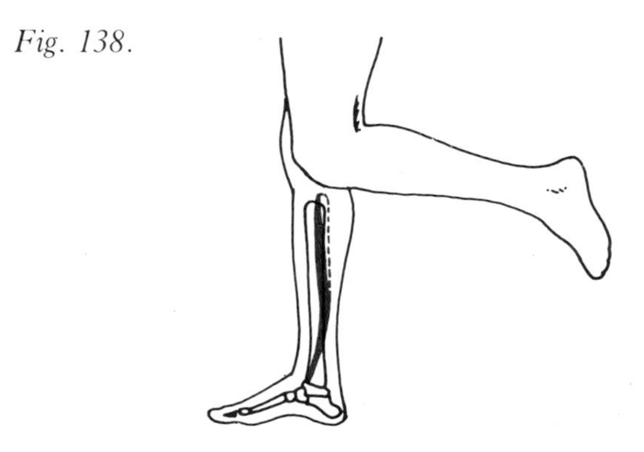

Fig. 138.

Chapter 4

ANATOMY AND FUNCTION OF THE TRUNK

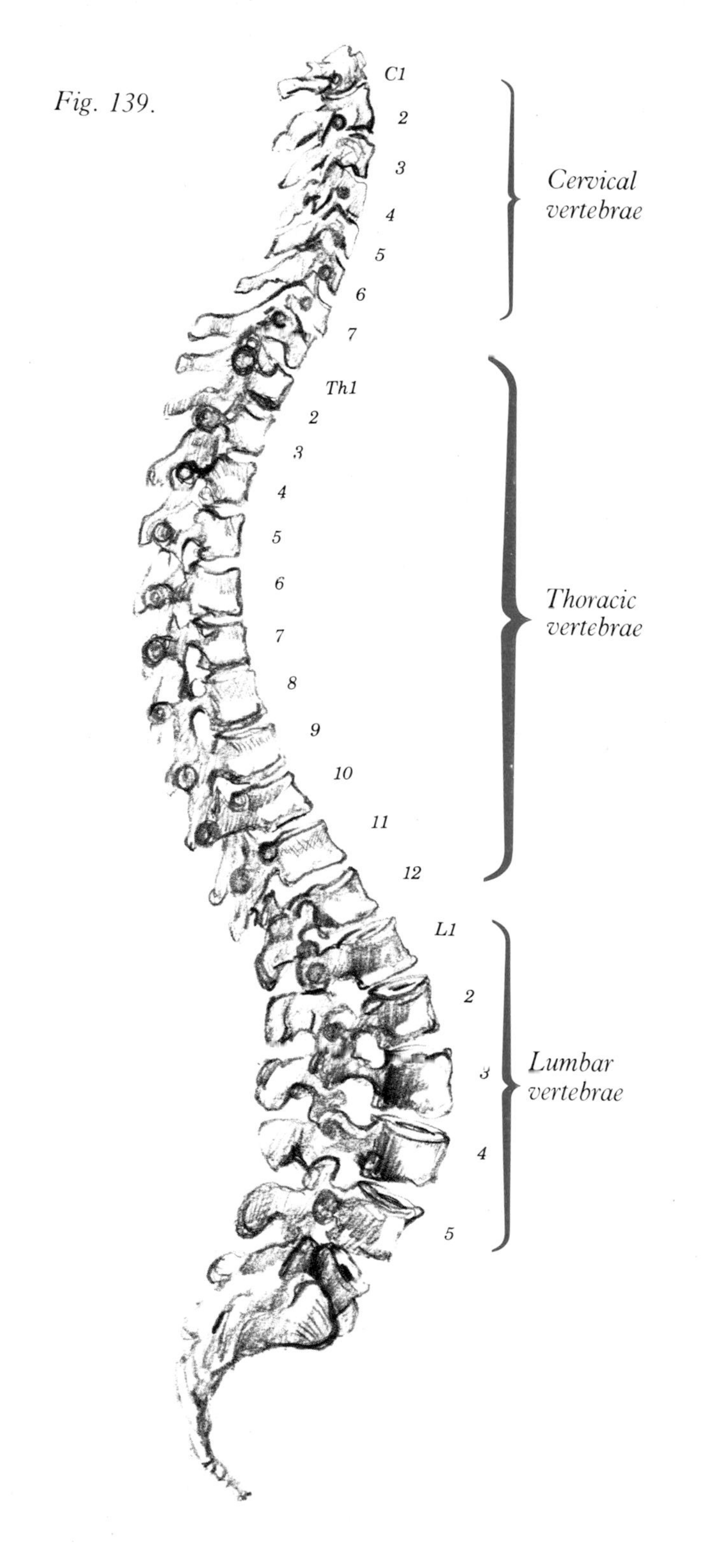

Fig. 139.

The vertebral column consists of seven cervical (of the neck) vertebrae which are numbered and usually written as C1, . . ., C7; 12 dorsal (or thoracic i.e. of the chest) vertebrae Th1, . . ., Th12; and five lumbar vertebrae L1, . . ., L5 (Figure 139).

Between the vertebrae there are intervertebral discs made up of an outer layer of fibrocartilage (annulus fibrosus) (1) and a more pulpy centre (nucleus pulposus) (2). Thus, the disc, with its tough fibrocartilage rim and soft, pulpy, highly elastic centre, acts as a shock absorber. Discs make up about one-third of the height of the vertebral column (Figure 140).

When two vertebrae are placed on top of each other, there is a substantial space between the stalks or pedicles of each vertebra; this is called the intervertebral foramen (3). It is through this space that the nerves pass from the spinal cord (4) and reach out to all the different parts of the body.

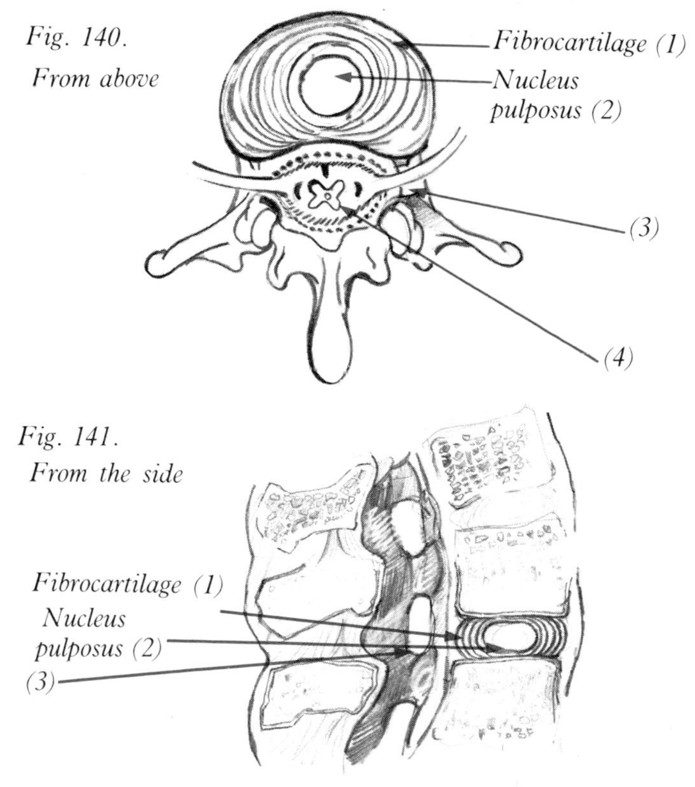

Fig. 140.
From above

Fig. 141.
From the side

Fig. 142.

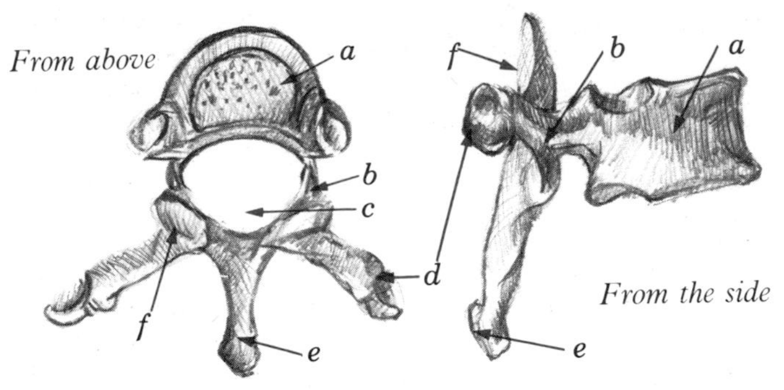

The following characteristics of a vertebra can be distinguished in Figure 142.

(a) Body.
(b) Neural arch.
(c) Vertebral opening.
(d) Transverse outgrowth or process.
(e) Spinous outgrowth or process.
(f) Articular outgrowth or process.

We are prevented from leaning too far back by the spinous processes and the tension of long ligament (1) which runs down the front of the spinal column (anterior longitudinal ligament) (Figure 143).

Fig. 143.

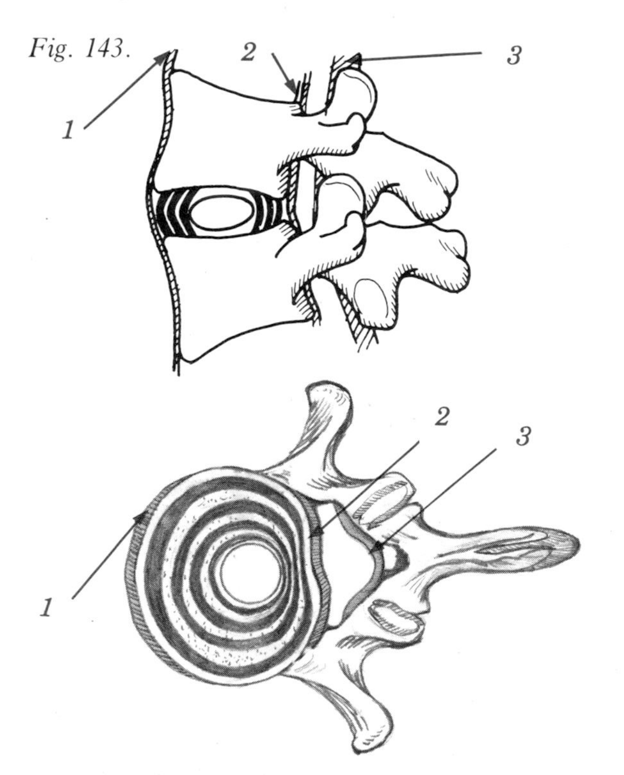

Forward bending is restricted partly by the back muscles, partly by the elastic ligament (3) which runs between the posterior parts of the neural arches (ligamentum flavum), and partly by the ligament which runs down the back of the vertebral bodies (i.e. in the anterior part of the vertebral opening). This ligament (2) is called the posterior longitudinal ligament.

The back complaints of athletes are often due to the spine being subjected to excessive or uneven stress, or the body being subjected to sudden movements while it is in an unfavourable position.

The back complaints of non-athletes are often due to poorly trained back muscles (even poorly trained muscles of the legs and abdomen), the wear and tear of lifting objects in a lopsided manner, sitting still, or working in a position where the body is tilted forward.

The pressure inside a disc varies according to the position of the body and external stress. The position which the body can assume to ensure that the least amount of pressure is placed on the lumbar spine is shown in Figure 144. Thus, lie on your back and support your lower legs on, say, a chair so that the iliopsoas does not pull (and thereby increase disc pressure) on your spine (see Figures 149 and 150). Sitting positions produce greater disc pressure than standing positions; a fact that is unknown to many!

Fig. 144.

This is because the back muscles must work harder (static work) when the body sits. Disc pressure is due to the bodyweight (mg) acting on the disc from above, and the contraction force (F) of the surrounding muscles (Figure 145).

The total pressing force in Figure 145 will be $mg + F$. Pressure (P) is calculated by dividing the force by the area of the disc. The L3 vertebra of an adult has an area of about 10 cm^2. The disc of a young person can withstand a stress of 800 kg, i.e. 8000 N. The ability of the discs of an older person to withstand stress is reduced to half of this. A young healthy disc can thus withstand a pressure of 800 N/cm^2. Compare these figures with those given on p.67.

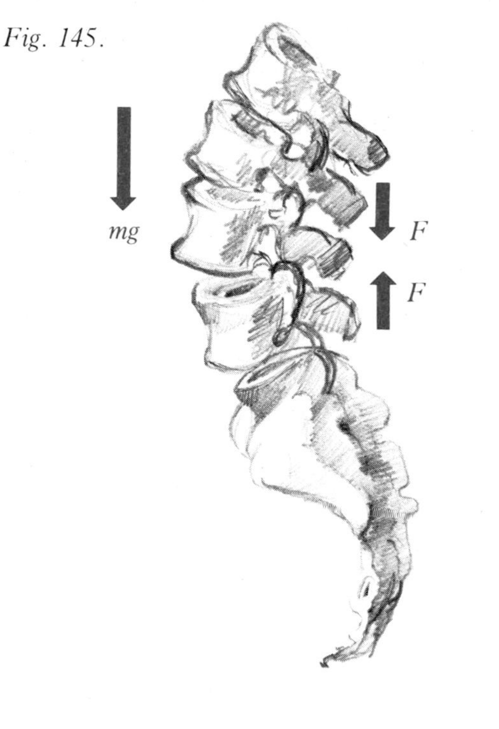

Fig. 145.

When a person carries something heavy the disc pressure is, of course, increased. Asymmetrical stress produces greater pressure than symmetrical stress (p.67). "Hanging" from a bar reduces the pressure on the discs and thereby unloads the back. At the same time the back muscles are stretched. For best results, "hang" with bent hips and support for your feet, (relaxed iliopsoas = straight back).

Fig. 146.

As a matter of curiosity it can be mentioned that if the stress is of long duration, or very great, a certain amount of fluid can be pressed out of the nucleus and we can actually measure a shortening of body height. A weight-lifter may be several centimetres shorter after a heavy training session.

If we measure disc thickness after a night's rest, we will find that they are thicker than if we measure them after a short period of normal activity. Therefore, we are somewhat taller in the morning than at the middle of the day.

We become shorter as we get older because the vertebral discs shrink as the tissues degenerate.

There are many reasons for back pain. A common reason is that the fibrous part of the disc tears and its nucleus is pressed backward, stretching the ligament which runs posteriorly along the vertebral bodies in towards the spinal canal. If the ligament stretches, pain is felt via its pain-sensitive cells. (There are only a very few nerves to register pain in the torn disc.) This type of pain can disappear if the back ceases to be stressed by the lifting of heavy objects, leaning forward while working, or sitting still. If the nucleus bulges too far out (Figure 148), it can press against the nerve root which passes through the intervertebral opening. Pain is then felt from the muscles that are supplied by this nerve. Thus, pain can be felt in the shoulder when a cervical disc is injured. Tense muscles, small vertebral displacements, or worn-down intervertebral cartilage can put the same pressure on the nerves, thereby causing pain. If the pain is felt in the leg, the name sciatica is used to describe it since the irritated nerve in this case is called the sciatic nerve.

Fig. 147.

Some of the nerves which pass out from a central spinal opening and into a muscle can be drawn out a little further by stretching the muscle. This can cause the nerve to be pressed against a protrusion on the disc, producing severe pain in the leg. One way to examine a person to see if his sciatic nerve is irritated, is to place him on his back and lift his leg as shown in Figure 147. This test is called the Lasègués test. This type of pain must not be confused with that felt by stiff people who try to stretch their hamstrings (p.55).

The fibrous part of the disc consists mainly of collagenous fibres, which stretch if they are subjected to prolonged stress. Hernia is commonly caused by working in a position which subjects the disc to such a force for a long period of time. Lifting heavy objects increases disc pressure which causes pressure in the nucleus pulposus (1) to rupture the fibrous annulus fibrosus (2).

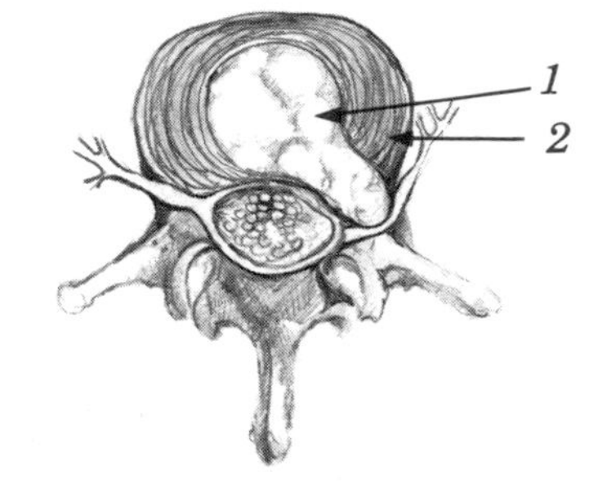

Fig. 148.

The increased pressure produced by simultaneously lifting something heavy and twisting the trunk (e.g. shovelling snow), is greatest in the posterior part of the disc which is not protected by extra ligaments (Figure 148). Such activities are therefore especially dangerous for those who suffer from back complaints.

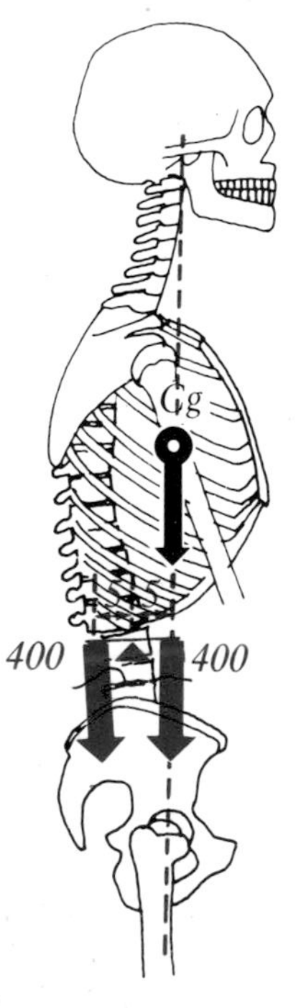

Fig. 149.
When a person stands, a line from his centre of gravity to the ground (vertical mid line) passes about 5cm in front of the centre of disc L3, and his back muscles lie about 5 cm behind it. Hence, the muscle force must equal 400 N in order to prevent his upper body from falling forward. The force acting on the disc is 400 + 400 = 800 N.

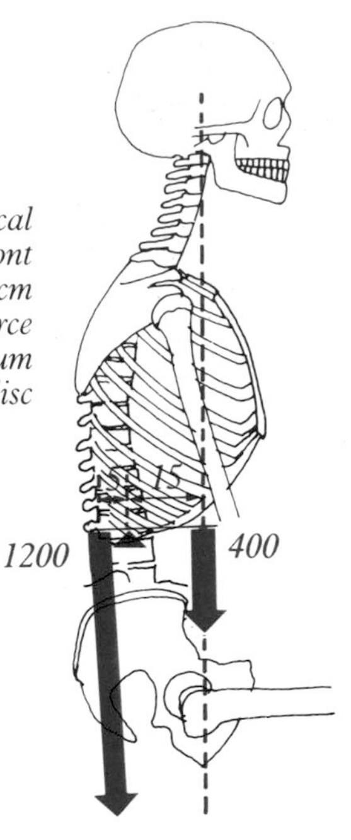

Fig. 150.
When a person sits, his vertical midline passes about 15 cm in front of L3. The muscle's lever arm is 5 cm (as in standing). Therefore, a force of 1200 N is required for equilibrium to exist. The force acting on the disc will be 1200 + 400 = 1600 N.

Using the moment of force law (p.32), we can calculate which stresses the spinal column is subjected to in different positions, lifts and training exercises. The person in the example below weighs 80 kg, of which 40 kg lies above L3. Distance is measured in centimetres and force is measured in newtons. Thus, 40 kg corresponds to 400 N.

When lifting an object it is important to stand so that the external lever arm (the distance from L3 to the point where the force of gravity acts on both the body and the object) is as short as possible. In the example below (Figure 151), the upper body (40 kg) + load (10 kg) is equal to 500 N. The back muscles are at work about 5 cm behind L3. When a person stands correctly (a), his external moment arm is about 20 cm, compared with 30 cm when he stands and lifts incorrectly (b). When he lifts from a sitting position (c), even very small weights subject his back to great work loads.

Fig. 151.

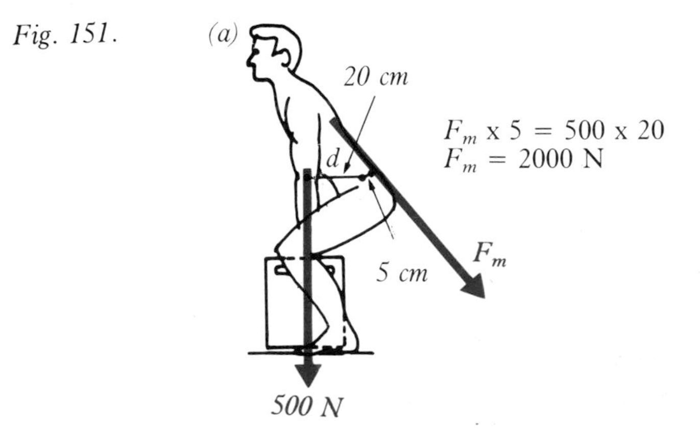

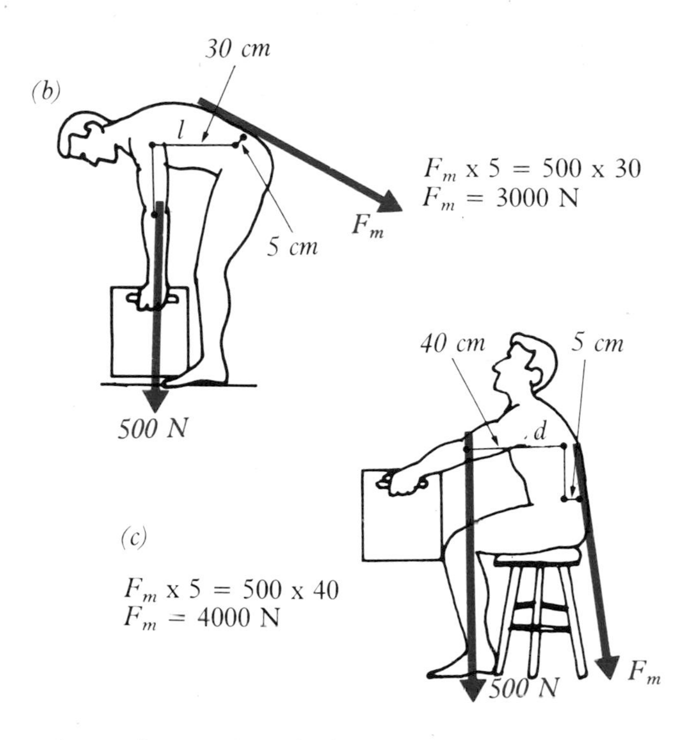

According to the calculations above, the pressure to which the disc is subjected by the muscles in (c), is half of what it can tolerate in a healthy condition (8000 N, p.64). This is true even when a person lifts a "mere" 10 kg from a sitting position. The discs are, however, unloaded when a person instinctively tenses his abdominal muscles plus his diaphragm (p.77). By doing so, he builds up the pressure in his abdomen which has a piston effect in an upward and downward direction (preventing collapse). The disc — which is a part of the abdominal cavity's back wall — is thus protected. Hence, the pressure produced by the back muscles can be reduced by about 40%.

Fig. 152.

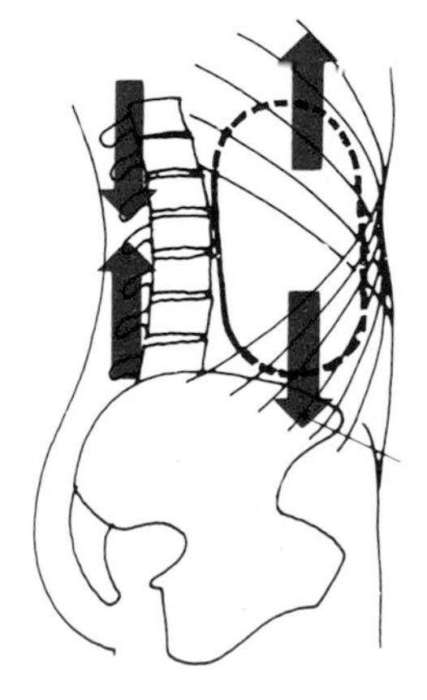

The calculations above show how important it is to have well-trained abdominal muscles to unload the back, and strong leg muscles in order to lift correctly, i.e. with bent knees.

If a load can be distributed symmetrically by carrying it with both hands, the stress to which the back is subjected is considerably reduced compared with carrying with only one hand.

In Figures 153 and 154, the body weighs 40 kg above L3. The load is 30 kg. The lever arm of the back muscles (for sideward bending) is 5 cm.

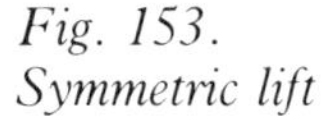

Fig. 153.
Symmetric lift

Total work load
150 N + 150 N + 400 N = 700 N
(the reader should also take account of the discussion accompanying Figure 145).

Fig. 154.
Asymmetric lift

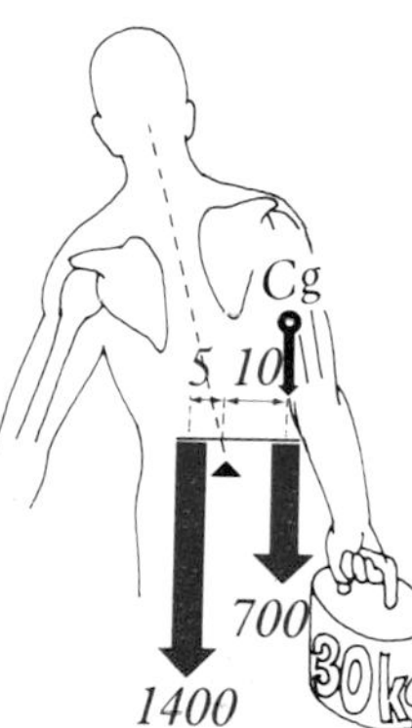

Suppose the centre of gravity of the body plus load falls 10 cm to the side of L3 (note that their common centre of gravity lies to the right of L3, even if the body tilts to the left). The back muscles must then contract with a force of $F_m \times 5 = 700 \times 10$. $F_m = 1400$ N. Total work load is 700 N + 1400 N = 2100 N.

The back is heavily loaded when the iliopsoas (p.45) is forced to work. The muscles are forced to work statically to lift the legs straight out in front of a person who hangs from a bar. If he weighs 80 kg his legs weigh 30 kg, the centre of gravity of his legs lies 40 cm from his hip joint, and his iliopsoas works with a lever arm of about 5 cm, then the force can be calculated as follows (Figure 155):

About 1900 N of this force strives to increase the sway of the lumbar spine. The force pressing the discs together will be about 2200 N. If the abdominal muscles do not keep the back straight, then a force of 2200 N could greatly increase the pressure in certain parts of the disc (Figure 156).

Fig. 155.

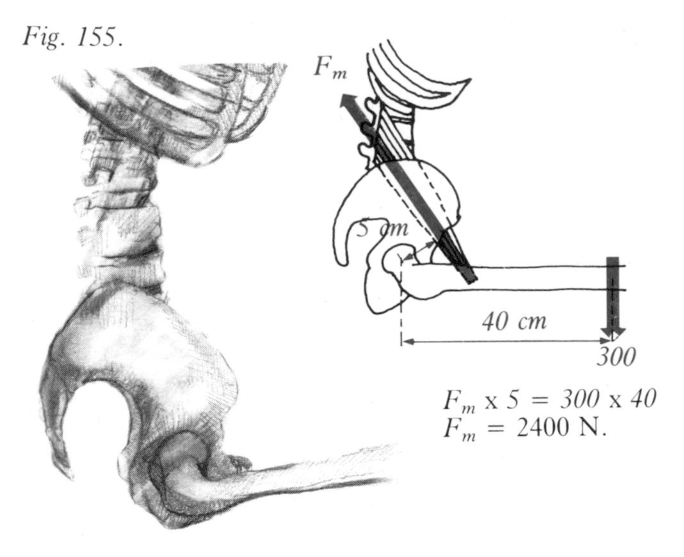

Such an exercise is thus unsuitable for people with weak abdominal muscles. Some sporting events, however, subject the iliopsoas to considerably greater stress.

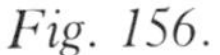

Fig. 156.

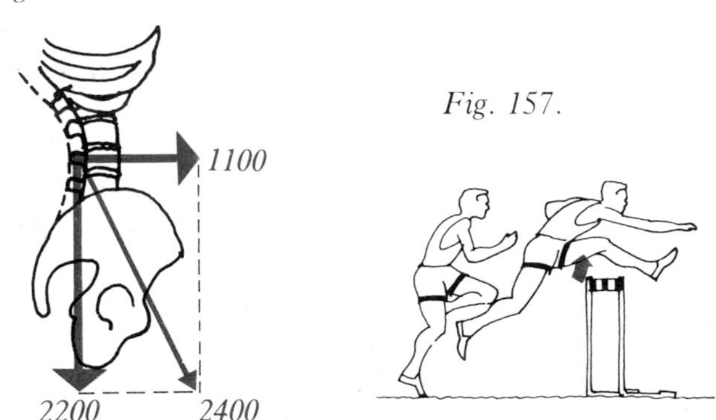

Fig. 157.

By way of comparison we can examine the forces that are necessary to swing a leg forward during running, jumping, hurdling, etc. The weight and acceleration (for which the iliopsoas is responsible) of the leg may cause the force of muscle contraction to reach about 4000 N.

When we consider exercises that engage the back muscles, we must also take account of the stabilising effect of the abdominal muscles. The diagrams below illustrate the different types of back and abdominal muscles found in the body.

Muscles of the back (erector spinae)

The muscles of the back can be classified schematically as follows.

Long back muscles (pass at least seven vertebrae).
Back muscles of average length (pass two to six vertebrae).
Short back muscles (go to the nearest vertebra).

The long muscles lie outermost (Figure 158). They are called:

(a) iliocostalis (from the hip bone to the ribs);
(b) longissimus (from the spinous outgrowths to the transverse outgrowths and ribs) and
(c) spinalis (between spinous outgrowths).

Fig. 158.

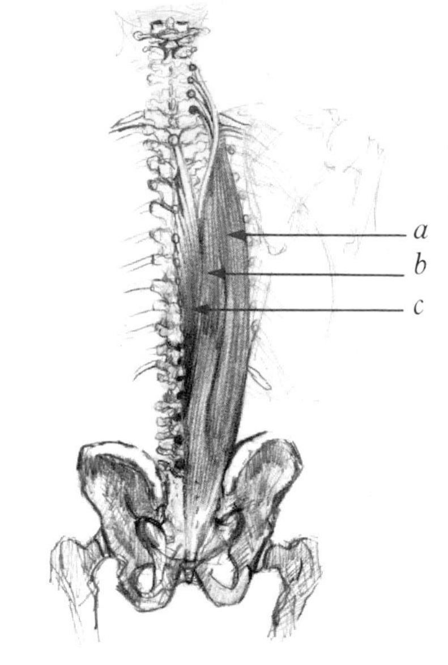

The back muscles of average length (Figure 159) are:

(d) semispinalis (passes four to seven vertebrae) and
(e) multifidus (passes two to three vertebrae).

Fig. 159.

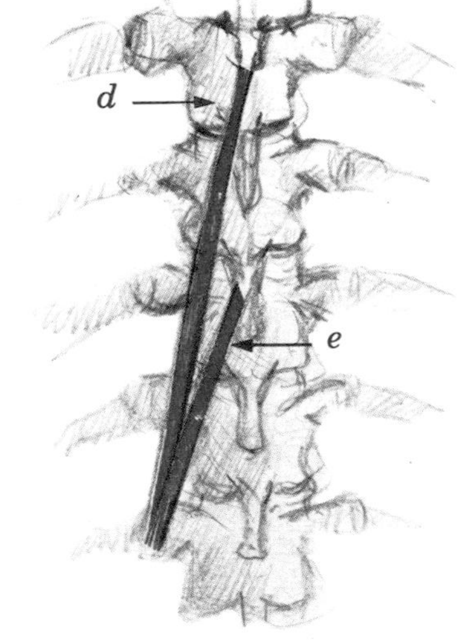

The short muscles passing from vertebra to vertebra (Figure 160) are:
(f) intertransversus (between transverse outgrowths);
(g) interspinales (between spinous outgrowths) and
(h) rotatores (between spinous and transverse outgrowths.

The muscles work together as a unit. The iliocostalis, however, is more adapted for taking part in sideward bending than the remaining muscles. The most important muscles for turning the trunk are the rotatores. A combination of bending and turning is important for all types of throwing.

Fig. 160.

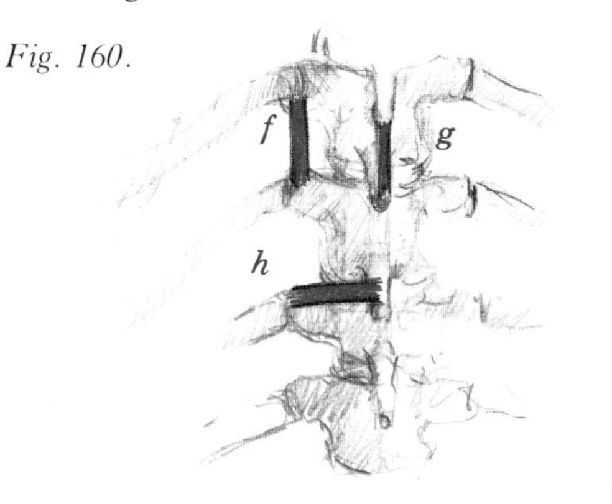

It is believed that cramps of the short muscles — in particular the rotatores — are the most common cause of so-called lumbago. If a cramp is brought about in one muscle, the muscles surrounding it contract to prevent movements that could tear it. This, in turn, cuts off the blood supply to the area, causing cramps in other muscles.

Fig. 161.

The cramp condition can be relieved by taking the load off the muscles (bedrest) and inducing relaxation (warmth, massage, muscle relaxing medicine). A cramp may result from overexertion, unaccustomed movements or minor vertebral displacements caused by sudden stresses. The best protection is well-functioning back and abdominal musculature.

Exercises for the back

In the following exercise the rotatores are forced to work. Stand on all fours; then, simultaneously lift your right arm and left leg (Figure 162). This activates the diagonal muscles which prevent the right shoulder and left hip from falling downwards. You should remain in that position for a few seconds so the muscles can work statically. Next, change positions by lifting your left arm and right leg.

Fig. 162.

Fig. 163.

Additional exercises for the back are given below, along with the effect they are considered to have.

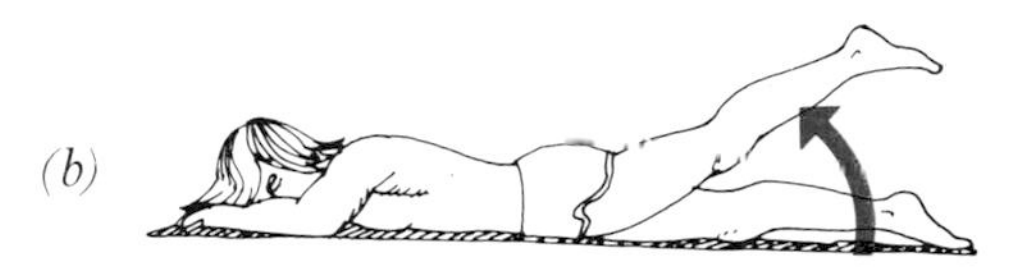

(a) This exercise should be done slowly. Well-trained young people can also turn sideways from the raised position in order to activate the rotatores. The work load to which the lower back is subjected can be increased by holding the arms stretched out in front of the body. Do not exaggerate backward bending of the neck. Do not go too far.

(b) Lift one leg at a time to train the hamstrings, the large buttock muscle and the lower back musculature. Do the exercise slowly. *Stop* in the raised position for static training. Do not combine exercises (a) and (b).

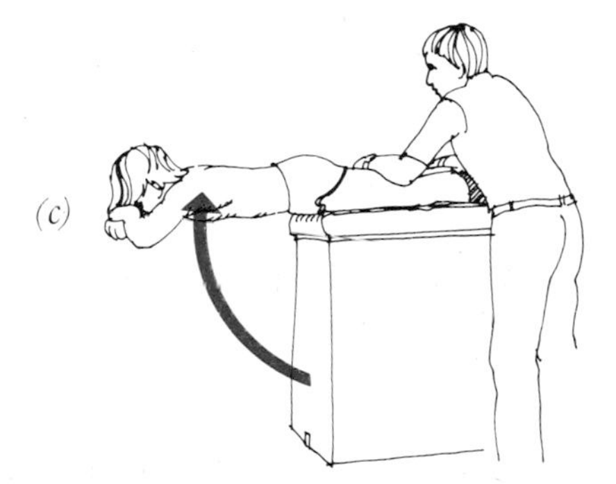

(c) Lie on a vaulting horse with your pelvis free as shown in the diagram. Keeping your back straight, pull yourself up until you reach the horizontal plane. You will be training all your back muscles statically; your hip extensors concentrically on the way up, and eccentrically on the way down. You can vary the work load with the help of your arms — little stress when your arms are by your sides, great stress when your arms are folded behind your neck. Light weights can be used.

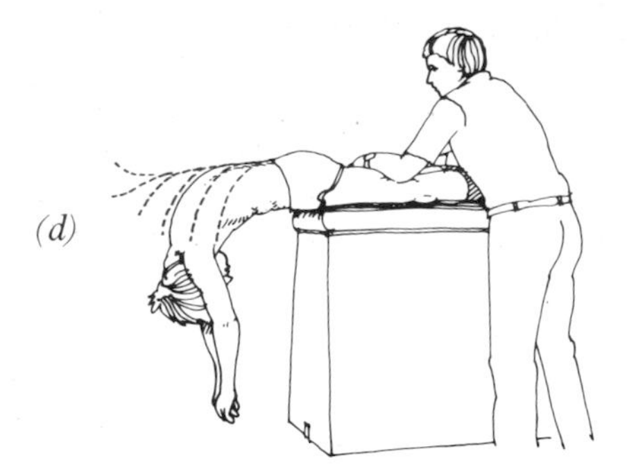

(d) As in (c) except that you should roll up to the horizontal plane one vertebra at a time beginning at the lumbar spine. Your back muscles are then forced to work concentrically (in the order L5, L4, . . ., C1) while rolling up and eccentrically while rolling down (C1, . . ., L5).

The diagrams below (Figure 164) show how the back muscles can be stretched by leaning forward without subjecting the back to stress, so that the whole musculature can relax.

Fig. 164.

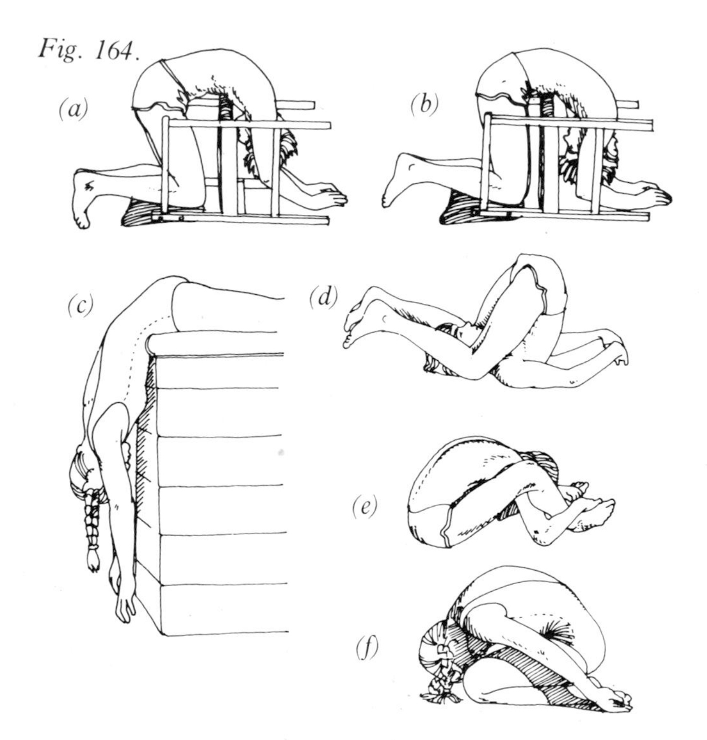

The above positions also reduce disc pressure by pulling on the back (traction).

When training the back muscles (Figure 165) one can (1) allow the trunk to fall forward — keeping the legs and back straight — until the horizontal plane is reached. Then (2) in an easy movement, bend further forward with flexed knees and back.

Fig. 165.

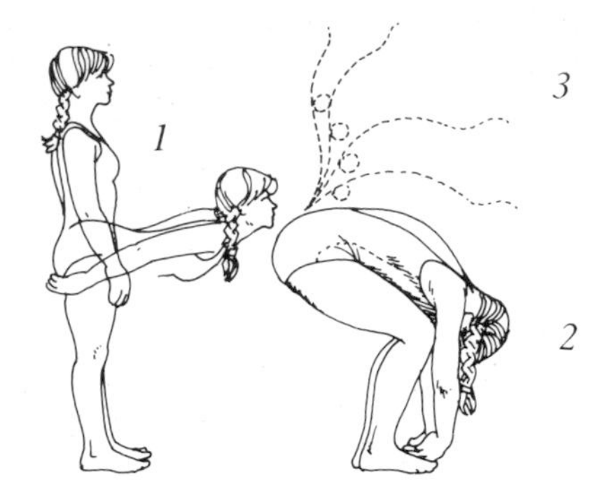

Next, (3) roll up one vertebra at a time, beginning at the lumbar spine. The first phase (1) stresses the muscles of the back statically. The work load increases as you approach the horizontal plane. You can increase the stress still further by making some "breast strokes" with your arms while you are in the horizontal plane. By doing so, you will be shifting the centre of gravity of your upper body even farther away from your hips. By falling forward beyond the horizontal plane, you can decrease the work load for your back muscles (the load will instead be borne by the posterior longitudinal ligament, p.64). People with back complaints and elderly people should avoid this.

Rolling up (3) forces the back muscles to work concentrically along the entire length of the spine [cf. Figure 163 (d)]. We are often told that rolling the trunk is a dangerous exercise, but there is no evidence that this is so. Moreover, trunk rolling exercises (Figure 166a) are often confused with hula-hula movements (Figure 166b).

Fig. 166.

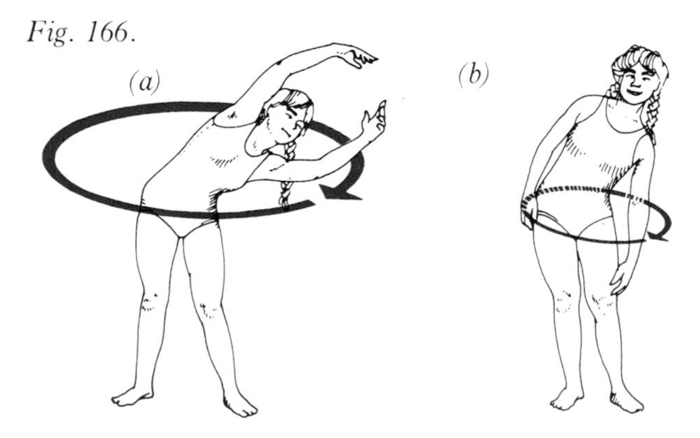

Trunk rolling can be done by moving the upper body (trunk) in large circles while keeping the hips and legs still. If this movement is performed when the body is warmed-up and if the pace is controlled, then there is no risk associated with it. By the same token, it can hardly be claimed to have especially beneficial effects.

Hula-hula movements increase the flexibility of the hip joint. Here, the feet and head are kept still while the hips are moved around in a circle.

As far as the neck is concerned, so-called head rolls should be avoided. There is a certain risk of doing these too quickly and thereby subjecting the ligaments and cartilage to wear and tear. Figure 167 shows the approximate range of movement of the head. The head cannot bend as far backwards as it can in all other directions. The shaded regions of the diagram mark the dangerous areas which the head may be forced to pass if it rolls around quickly. However, if you roll slowly and under control, you may follow the outer limits of the cervical spine's range of motion without any problems.

Fig. 167.

Abdominal muscles

As we have seen when examining movements that strain the back, a well-functioning abdominal musculature unloads the back during lifting, and stabilises the spinal column (i.e. the abdominal muscles are antagonists to the back muscles). The examples given on p 66 show that the back muscles are always trained during lifting, standing, sitting, etc. The majority of people have abdominal muscles that are too weak in relation to their back muscles. Thus, general abdominal muscle training (Figure 168a) can be recommended for everyone. Athletes should strengthen their abdominal muscles with exercises which subject them to a heavy work load (Figure 168b). After that they should train their hip flexors (iliopsoas, rectus femoris) and back extensors for strength. (For such exercises, see p.45 and 69.)

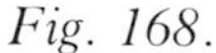

Fig. 168.

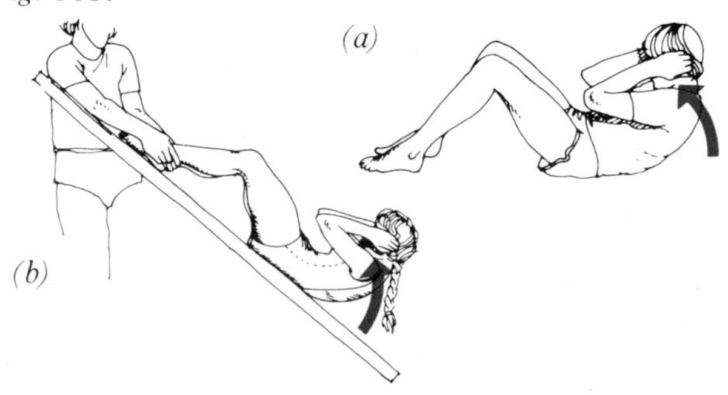

In order to judge correctly which exercises to choose and how to execute them, we must first understand how the abdominal muscles function. It is also important to familiarise ourselves with the connection which exists between the back, abdomen and hip flexors.

There are four different kinds of abdominal muscle.

Fig. 169.

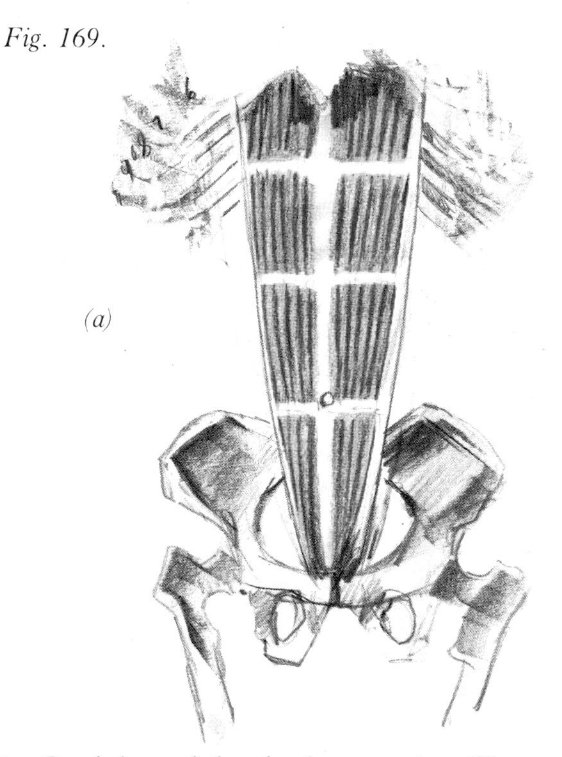

(a) Straight abdominal muscle The rectus abdominis originates from the apex of the breast bone and is inserted into the upper part of the pubic part of the hip bone. When it contracts the body bends forward at the lumbar and thoracic spines. If you lie on your back [as in Figure 168 (a)] and pull your upper body up as far as possible without tipping your pelvis forward (no movement at the hip joint), then you will maximally shorten your straight abdominal muscle.

(b) External oblique abdominal muscle The external oblique abdominal muscle arises from the anterior inferior part of the chest (lower ribs). This muscle becomes a tendinous expansion (or aponeurosis) which covers the straight abdominal muscle above, and below it is inserted into the crest of the hip bone and groin ligament. In the lower part of the abdomen the flat tendons of the two sides intersect.

Fig. 170.

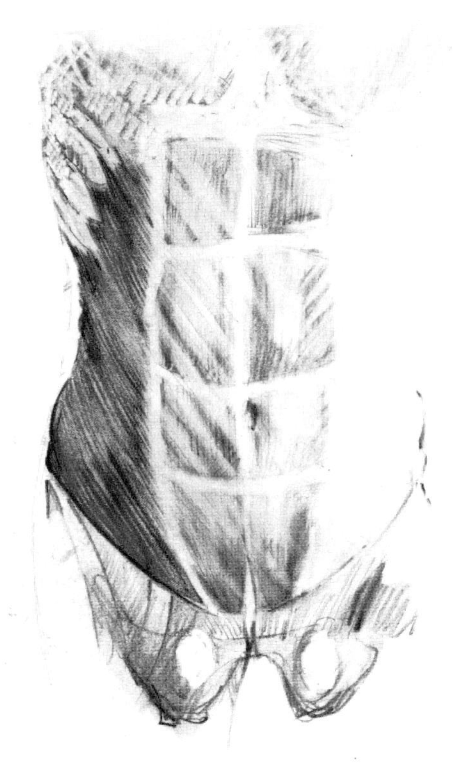

Fig. 171.

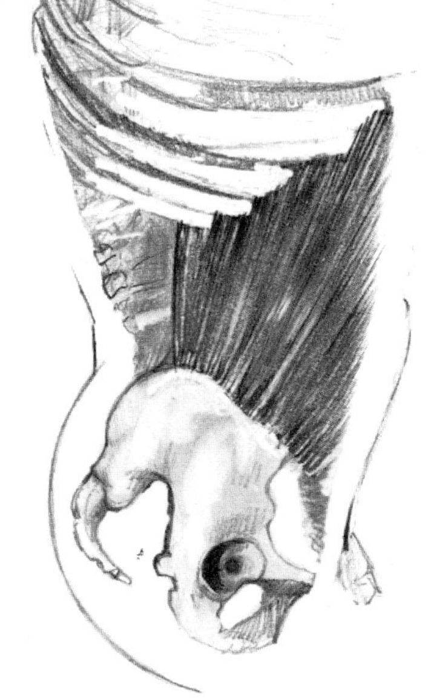

(c) Internal oblique abdominal muscle The internal oblique arises from the hip bone and groin ligament, becomes a tendinous expansion passing under the straight abdominal muscle, and is inserted into the fibrous band which runs down between the pair of straight abdominal muscles.

Fig. 172.

The oblique abdominal muscles act to assist the straight muscle. In addition, they can turn the trunk. In diagonal sit-ups (Figure 172) the stress to which the oblique abdominal muscles are subjected will increase so that if the right shoulder is turned towards the left hip, the right external and left internal oblique muscles will be engaged.

Fig. 173.

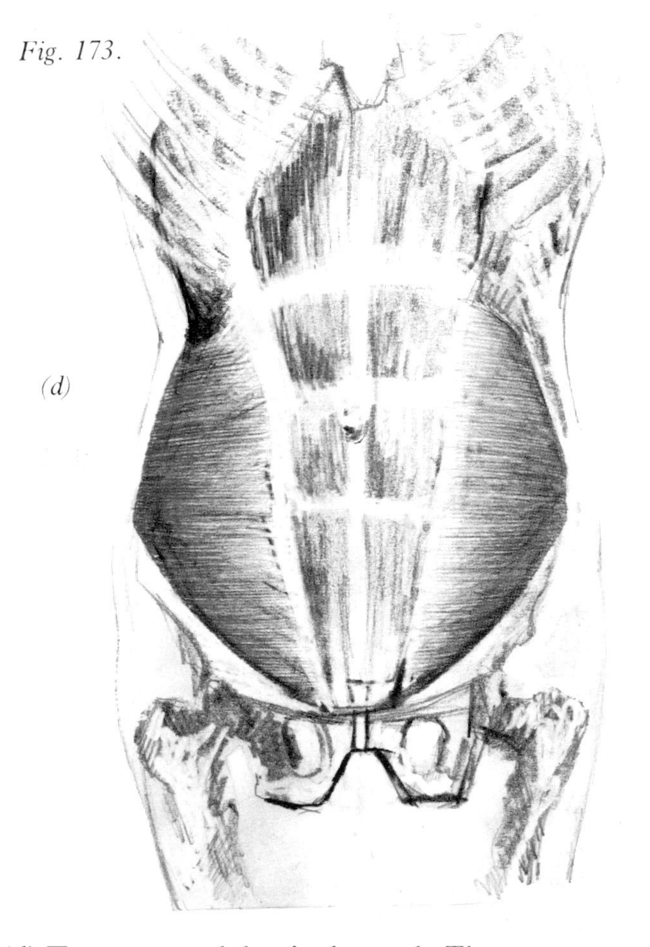

(d) Transverse abdominal muscle The transversus abdominis is not connected with any movements. It only affects the figure (by pulling the abdomen in). It is also brought into action when the abdominal pressure is increased by so-called abdominal presses. All the abdominal muscles can, upon contraction, increase abdominal pressure (which would try to widen the abdominal cavity). In this way, the discs are unloaded during lifting (p.67).

Fig. 174.

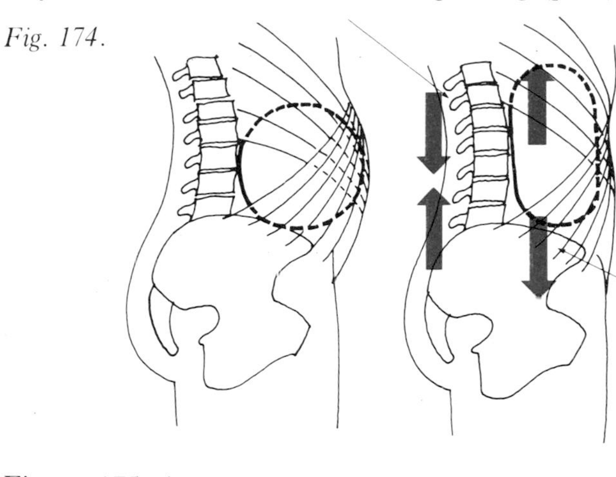

Figure 175 shows a horizontal cross section of the body. From it we can detect the relationship between the back and abdominal muscles.

Fig. 175.

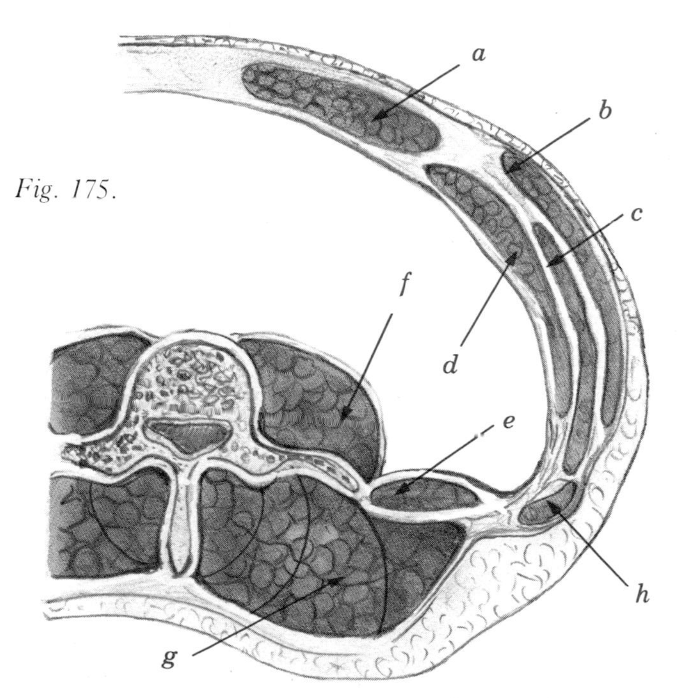

(a) Straight abdominal muscle (rectus abdominis)
(b) External oblique abdominal muscle
(c) Internal oblique abdominal muscle
(d) Transverse abdominal muscle
(e) Square lumbar muscle (quadratus lumborum)
(f) Iliopsoas
(g) Back extensors (erector spinae)
(h) Broad back muscle (latissimus dorsi, p.84)

Quadratus lumborum

Fig. 176.

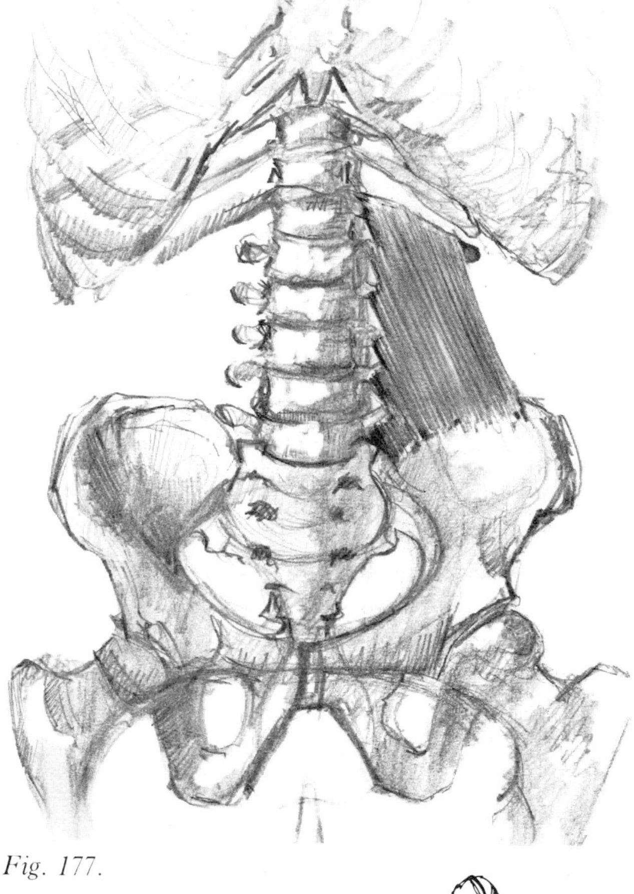

Fig. 177.

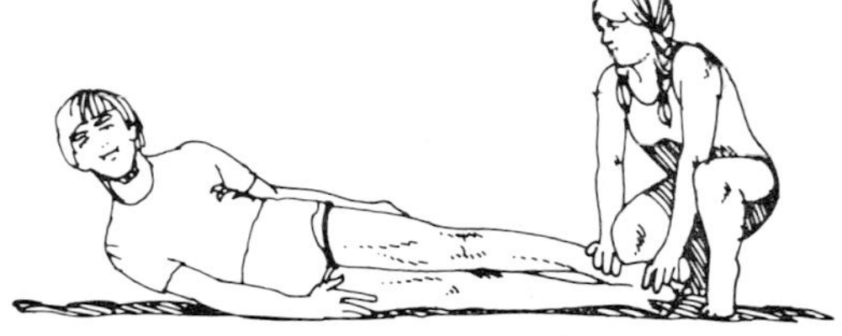

Here, we should also mention the most important sideward flexor. Its name is quadratus lumborum. It takes part in sideward bending when the movement occurs at the lumbar spine. Compare this with the movements described on p.40 which take place at the hip and also engage the abductor muscles.

Fig. 178.

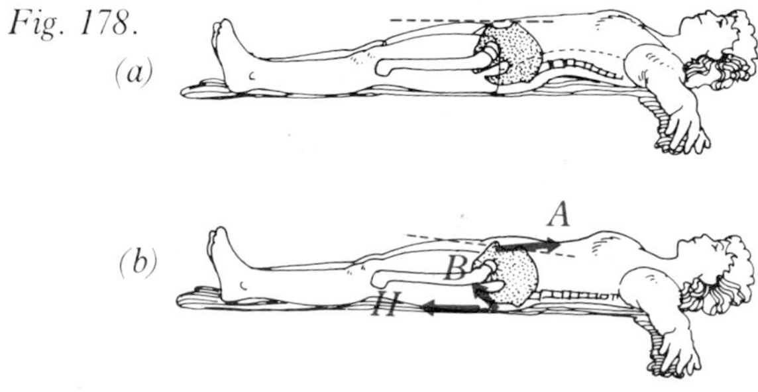

When a person lies on their back, there is a certain curvature on the lumbar spine which is due to the natural curvature of the spine as well as tension in the iliopsoas. The back can be pressed to the floor [Figure 178 (b)] by using the abdominal muscles (A), buttock muscles (B), and hamstrings (H) to tip the pelvis backwards. This would be found easier to do if the knees were bent slightly and the head lifted a little. Thus, the iliopsoas has been relaxed and a forward bending of the spine initiated.

Fig. 179.

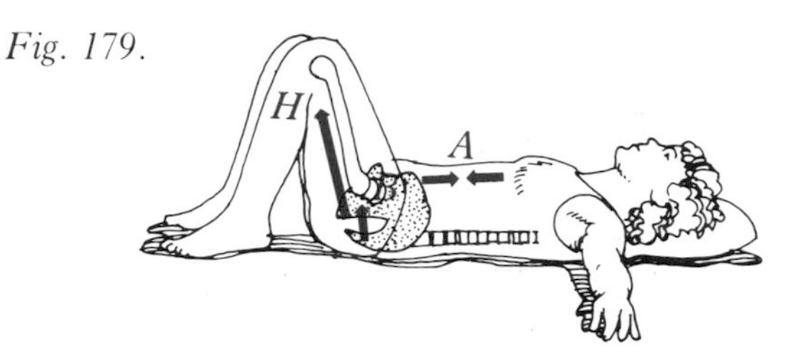

If you lie on your back and lift your legs, keeping them straight, you will easily reach a position where your back curvature is considerably greater than it is in a standing position. This is because the force with which the iliopsoas lifts the legs also acts on the back in such a way that the curvature is increased (see p.68).

Fig. 180.

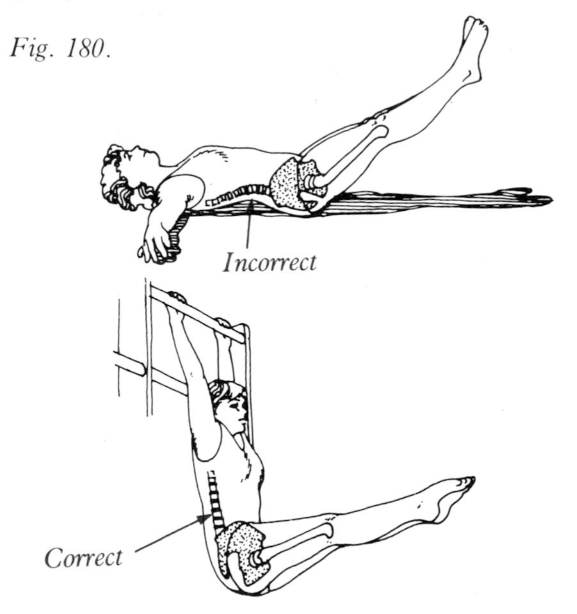

Exercises which require a person to lift straight legs, or give resistance to hip flexion (p.44), should only be undertaken if the abdominal muscles are capable of stabilising the back (tip the pelvis). Such exercises should be left to those who are relatively well-trained. Their aim, strictly speaking, is to strengthen the hip flexors. Nevertheless, there is a static training effect from which the abdominal muscles benefit.

If a person wants to train only their abdominal muscles, they should make sure that movement takes place at the spine and not at the hips. The most common exercise for strengthening the abdominal muscles is the sit-up. The best way to prevent a sit-up from procurring the aid of the iliopsoas is to be so bent at the hip that the muscle cannot contract with any degree of force. A common mistake is to bend the knees (with accompanying hip bending) at the same time as the feet are held under a wall-bar. In this position, the hip is not bent enough to disengage the iliopsoas and, because the feet are supported, the body can sit up without any great effort on the part of the abdominal muscles. If you assume the position shown in Figure 181 — without supporting your feet — you must roll up vertebra by vertebra (i.e. engage your abdominal muscles) before you can bend your hip and thereby pull yourself up.

Fig. 181.

The best way to train your abdominal muscles is by strongly flexing your hip, making sure that your feet are unsupported. It is then only possible to do half a sit-up (i.e. maximal contraction of your abdominal muscles) without bending your hips.

Fig. 182.

If you remain in the raised position and rotate to each side, you will be training your straight abdominal muscle statically as well as increasing the work load for your oblique abdominal muscles. You can vary the quantity of exercise by (1) keeping your arms straight by your sides (easy), (2) crossing your arms over your chest or (3) placing your hands over your ears (difficult) (Figure 182). The work load can be further increased by adding weights.

When we consider how the external stress increases during a sit-up (Figure 183), we see that the centre of gravity of the upper body lies some way from the hip joint at the beginning of the exercise, and that this distance decreases as the body rolls up vertebra by vertebra [Figure 183 (a)]. External stress (F_1) decreases as the abdominal muscles contract.

If a person lies on a sloping plane, the external stress will increase as he pulls himself up and as his abdominal muscles shorten (i.e. weaken). He would thus find this exercise very demanding [Figure 183 (b)].

Fig. 183.

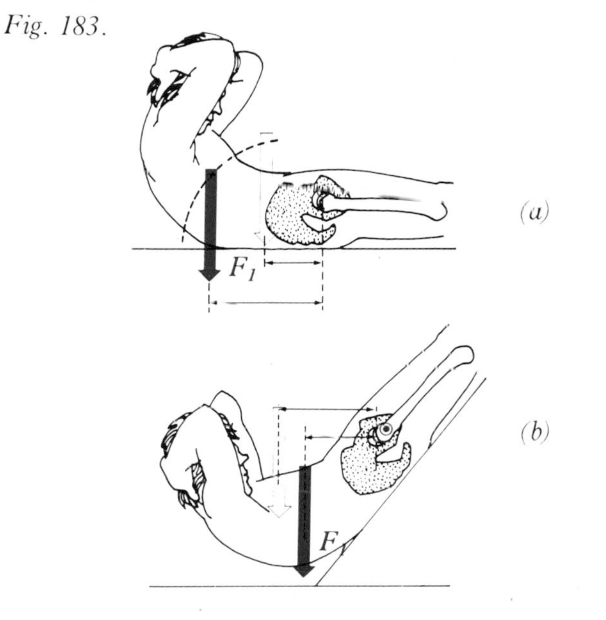

Which of the following is the most demanding: (1) sitting up on a slanting bench (45° slope) or (2) sitting up from the floor with 10 kg on your chest? Try them!

A person exercising on a slanting plane must have support for his feet. This means that he is tempted to use his iliopsoas at an early stage instead of rolling one vertebra at a time (i.e. working his abdominal muscles).

The four examples below are designed to illustrate the relationship between external forces (the weight of the legs and upper body) and internal forces (iliopsoas and abdominal muscles).

Fig. 184.

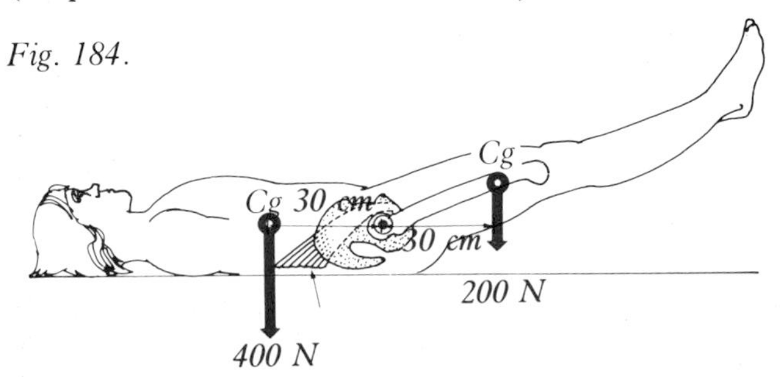

(1) A person who weighs 60 kg lies down as shown in Figure 184. The centre of gravity of his upper body, which weighs about 40 kg, is presumed to lie 30 cm above his hip joint. His legs weigh 20 kg. Their centre of gravity lies 30 cm below the hip. Thus, according to the formula, $M = F \times d$:

The moment of force of the upper body,

$$M = 400 \times 30 = 12000 \text{ Ncm}.$$

The moment of force of the legs,

$$M = 200 \times 30 = 6000 \text{ Ncm}.$$

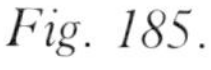

Fig. 185.

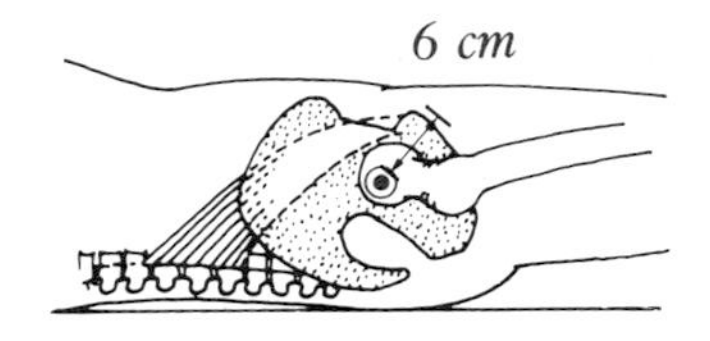

(2) The iliopsoas (the main muscle group responsible for hip flexion) is presumed to be positioned so that its direction of pull in relation to the hip is 6 cm. If its contraction force reaches 1000 N, then the moment of force acting on the leg will be 1000 x 6 = 6000 Ncm (i.e. equal to the moment of force of the legs, according to Figure 184). He can thus begin to lift his legs. If the muscle force is even greater, he can lift his legs with greater speed.

Fig. 186.

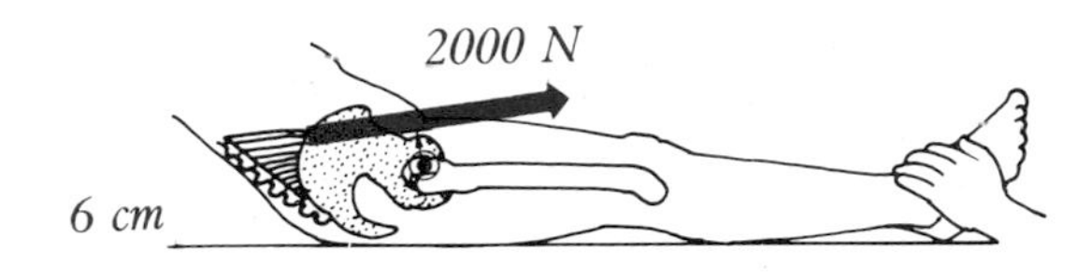

(3) If his legs are fixed and the iliopsoas contracts with a force of 2000 N, then the muscle's moment of force will be 2000 x 6 = 12000 Ncm and he can lift his upper body (as shown in Figure 186).

Fig. 187.

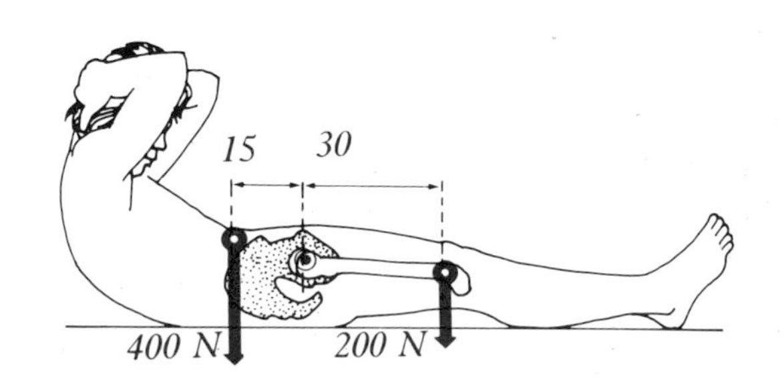

(4) In order to sit up without supporting his legs, he must roll up one vertebra at a time using his abdominal muscles until the centre of gravity of his upper body lies 15 cm (instead of 30 cm) from his hip. The iliopsoas will now lift both his upper body and his legs. By bending a little further forward at the spine, he can sit up without moving his legs. If his legs are bent, and they are unsupported, his abdominal muscles will be forced to work over an even greater distance before his hips can be bent.

The diagrams below (Figure 188) present some variations of abdominal muscle training. You should make sure that movement takes place at your spine (easy forward bending) before possible movements take place at your hip joint. Never subject your legs to more stress than your abdominal musculature (whose job it is to hold your back slightly bent forward) can cope with.

Fig. 188.

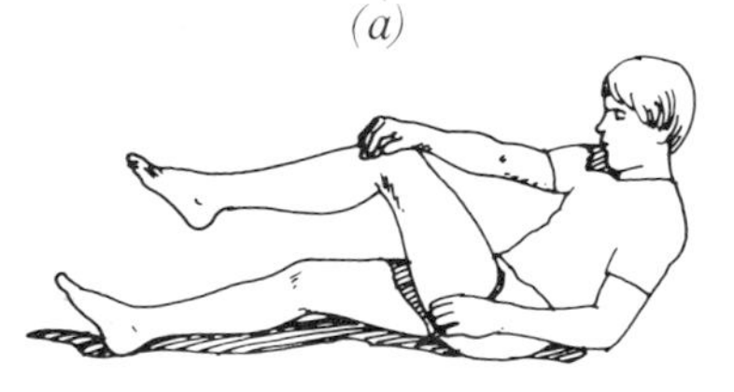

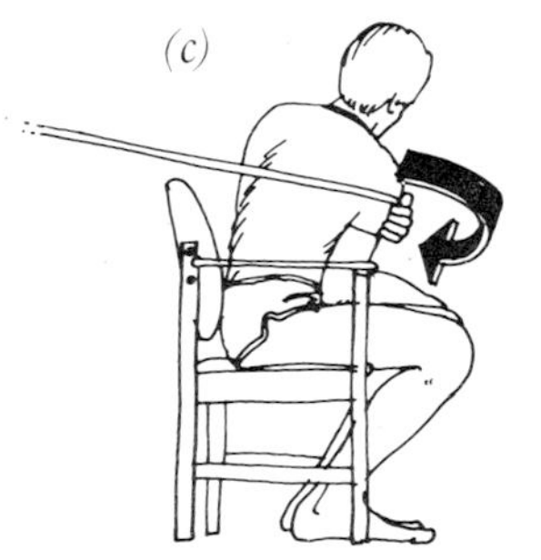

There is rarely ever a need of stretching exercises for the abdominal muscles.

Respiratory muscles

Here we will consider the structure of the rib cage and function of the breathing musculature in outline only. The reader is referred to the literature on physiology for a deeper study.

The rib cage consists of 12 ribs (or costae). The first 10 (from the top) are attached directly (first seven) or indirectly to the breast bone (body of the sternum) via costal cartilages. Ribs numbers 11 and 12 have no anterior attachment and are therefore called floating ribs.

Fig. 189.

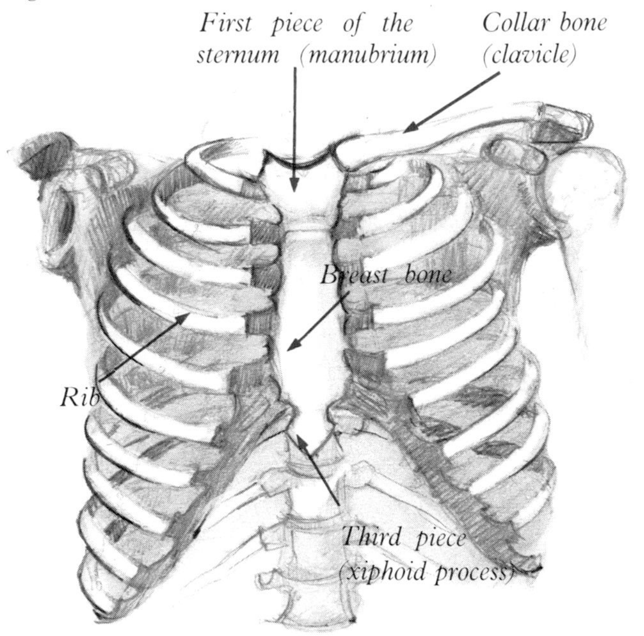

The chest and abdominal cavities are separated by the most important muscle of respiration, the diaphragm. This muscle arises from the lumbar vertebrae, from the lower ribs, and from the xiphoid process which is at the bottom of the sternum. It arches up like a dome into the chest cavity (Figure 190). When the muscle fibres (2) contract, they become less arched and thus cause the central tendinous part (1) of the dome to descend. During contraction, the volume of the chest cavity increases (inspiration) and the volume of the abdominal cavity decreases. The abdomen thus bulges out, which is why this type of breathing is called abdominal breathing (deep breathing).

Fig. 190.

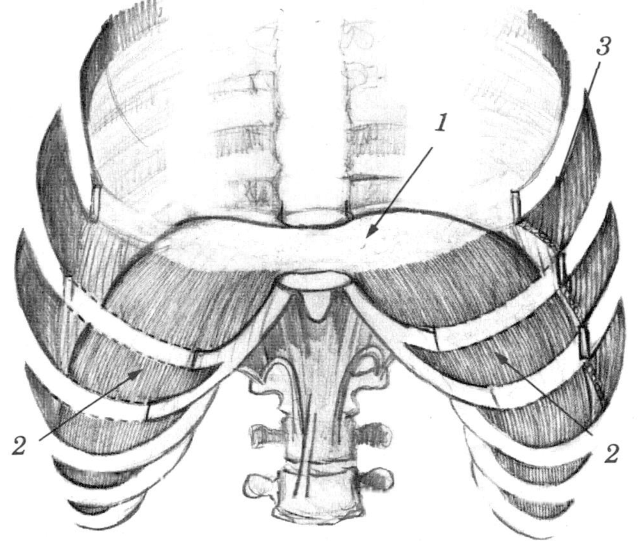

Beside functioning as a breathing muscle, the diaphragm assists the abdominal muscles in increasing intra-abdominal pressure. This is important when it comes to lifting heavy objects (see p.73). The volume of the chest cavity can also be increased by raising the ribs. This is accomplished by the action of the muscles which are attached to the ribs. The direction of the muscle fibres is such that they lift the lower ribs up towards the upper ribs (3). These muscles are called external intercostal muscles. This latter type of breathing is called chest breathing. A number of muscles (back, chest, neck, etc.) may exert an influence on the rib cage during forced breathing.

Chapter 5

ANATOMY AND FUNCTION OF THE ARM

The movements of the arm are controlled by many muscles. Three different muscle groups are responsible for the movements of the shoulder joint.

Fig. 191. A

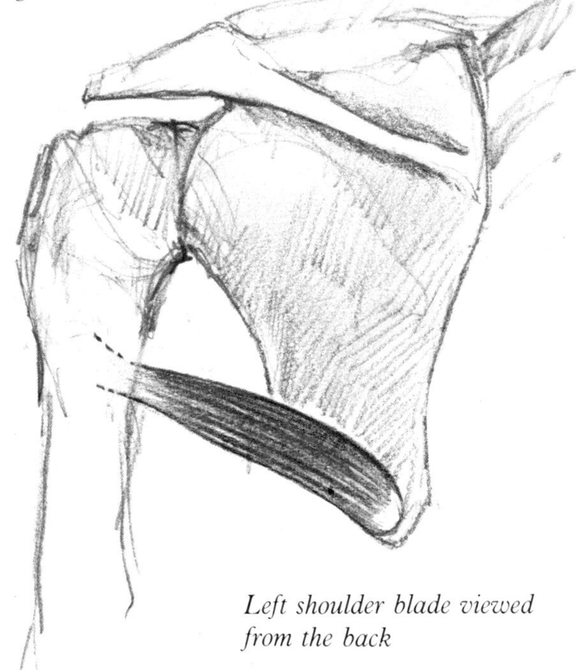

Left shoulder blade viewed from the back

B

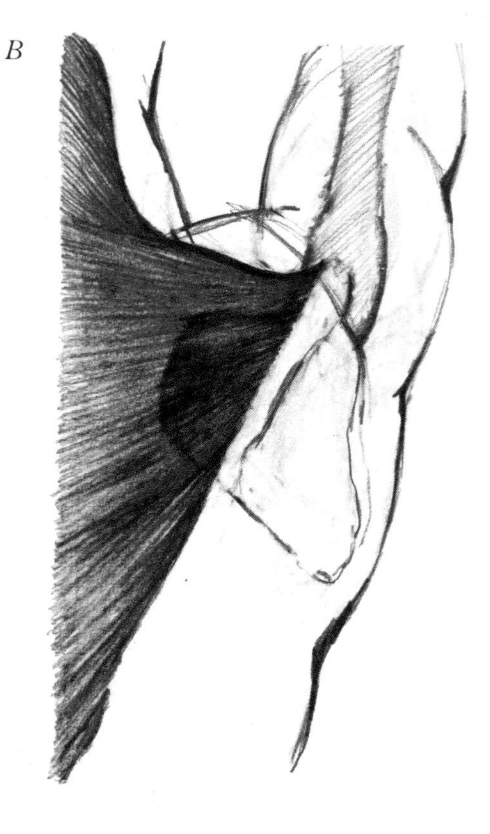

Group A. Muscles which have their origin on the shoulder blade and which are inserted into the upper arm.

Group B. Muscles which have their origin on the trunk and their insertion on the shoulder blade.

Group C. Muscles which have their origin on the trunk and their insertion on the arm.

The shoulder blade (scapula) is formed like a triangle. A sharp spine (1) (spine of the scapula) projects backward and forms two depressions on the posterior surface of the shoulder blade. The outer end of the spine forms a flat roof (2) (acromion) overhanging the shoulder joint. The anterior part of this roof articulates with the collar bone (3) (clavicle). This joint is the outer collar bone joint (4) (acromio-clavicular joint).

The inner collar bone joint (5) (sterno-clavicular joint) articulates with the manubrium of the breast bone (sternum) (6). The outermost part of the shoulder blade presents a shallow articular surface

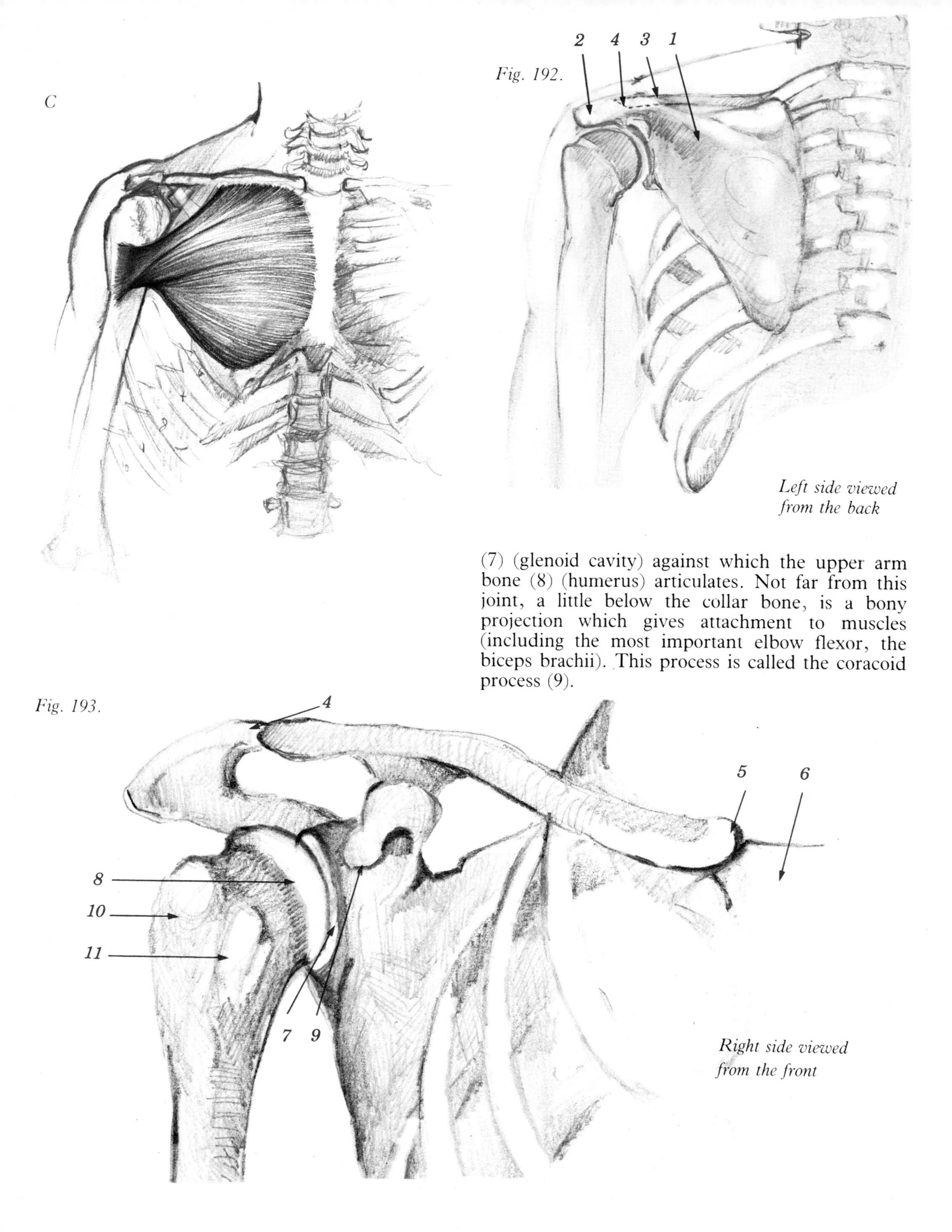

Fig. 192.

Left side viewed from the back

Fig. 193.

Right side viewed from the front

(7) (glenoid cavity) against which the upper arm bone (8) (humerus) articulates. Not far from this joint, a little below the collar bone, is a bony projection which gives attachment to muscles (including the most important elbow flexor, the biceps brachii). This process is called the coracoid process (9).

The arm articulates with the shoulder blade at a ball and socket joint, thus allowing movement in all planes.

(1) Forward and backward swings (flexion–extension).
(2) Outward and inward swings (abduction–adduction).
(3) Outward and inward turns (lateral and medial rotation).

Practically the whole of the shoulder blade functions as an area of origin for muscles. Between the neck and the shaft of the upper arm bone, we can distinguish two roughened projections of bone from which slight ridges run continuous with the shaft. The outer projection (10) (greater tuberosity) and the anterior projection (11) (lesser tuberosity), together with their ridges, will be referred to here simply as the external and anterior areas of attachment.

Many of the muscles of the shoulder blade have only Latin names. Translations of certain parts of the Latin words are therefore given in parentheses.

Group A

Group A consists of the following five muscles.

Fig. 194.

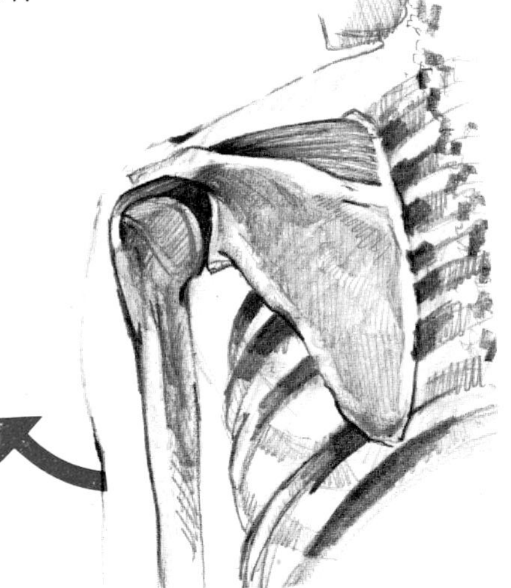

Left side viewed from the back

(1) Supraspinatus (supra = above, spina = spine) lifts the arm. It lies below the large muscle which give the shoulder its contour, the deltoid muscle (p.83) (Figure 194).

Fig. 195.

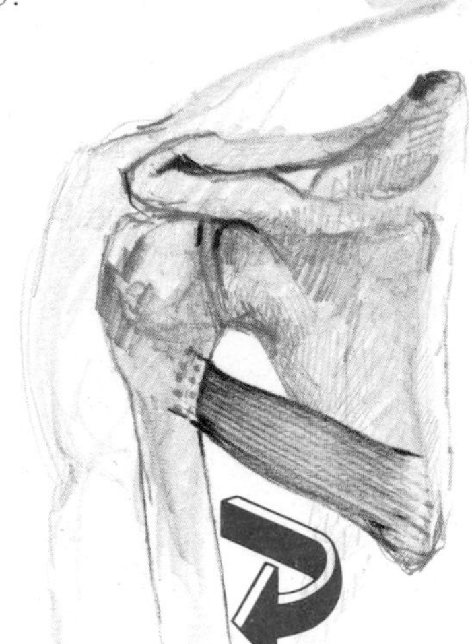

Left side viewed from the back

(2) Teres major (teres = rounded, major = greater). This large rounded muscle brings the arm to the side (adducts) and rotates it inwards. It works with the latissimus dorsi muscle (p.84) (Figure 195).

(3) Infraspinatus **(4)** Teres minor

In the gap between the muscles described above, there are two muscles which together adduct the arm and rotate it outwards. These muscles are called infraspinatus (infra = below) and teres minor (minor = less, smaller) (Figure 196).

Fig. 196.

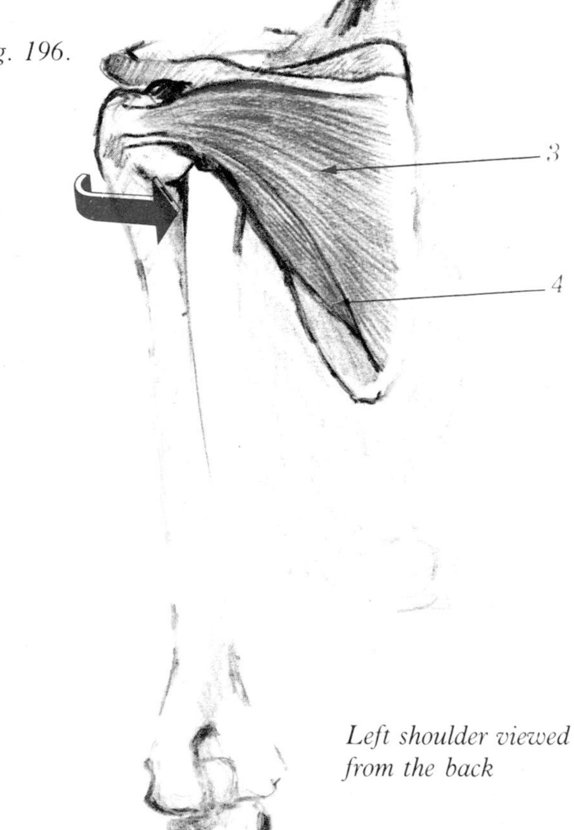

Left shoulder viewed from the back

(**5**) Subscapularis

Covering the inside of the shoulder blade (against the back wall of the rib cage) is a muscle which adducts the arm and rotates it inwards. Its name is subscapularis (sub = under, scapula = shoulder blade) (Figure 197).

Fig. 197.

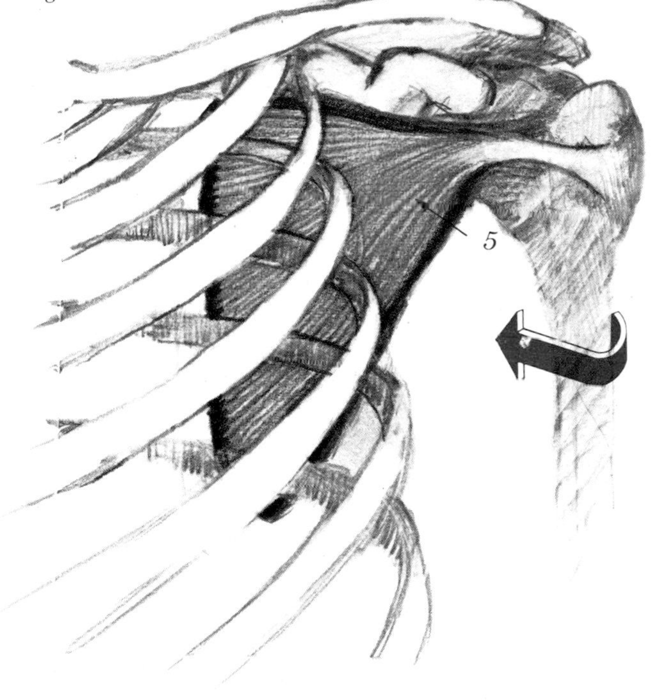

Left shoulder viewed from the front

Fig. 198.

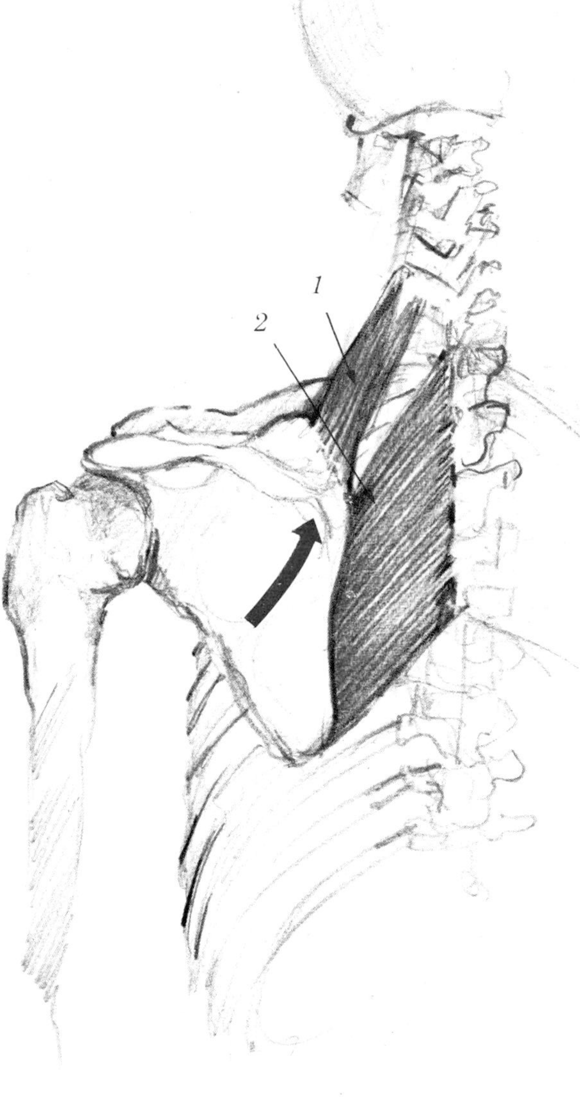

Group B

In order to move the arm with sufficient force, the articular surfaces of the shoulder blade must be positioned in such a way that the arm has an optimal starting point for its movement. The shoulder blade can be: (1) elevated–depressed (10-12 cm), (2) abducted – adducted (15 cm) and (3) rotated outwards and inwards.

Outward rotation means that the articular surfaces of the shoulder blade are directed outward and upward (Fig. 200).

The muscles which raise the shoulder blade are:

1. Levator scapulae (levator = "raiser")
2. Rhomboids (major and minor)

Lifting and a certain inward rotation occur simultaneously (Figure 198).

Both these muscles above are covered by trapezius.

3. Trapezius

Origin: Base of the skull and the spines of the cervical and thoracic vertebrae.

Insertion: Spine of the shoulder blade and external part of the collar bone.

Function: Raises and adducts the shoulder blade and rotates it outwards. Turns the head and bends the neck backwards (Figure 199).

Fig. 199.

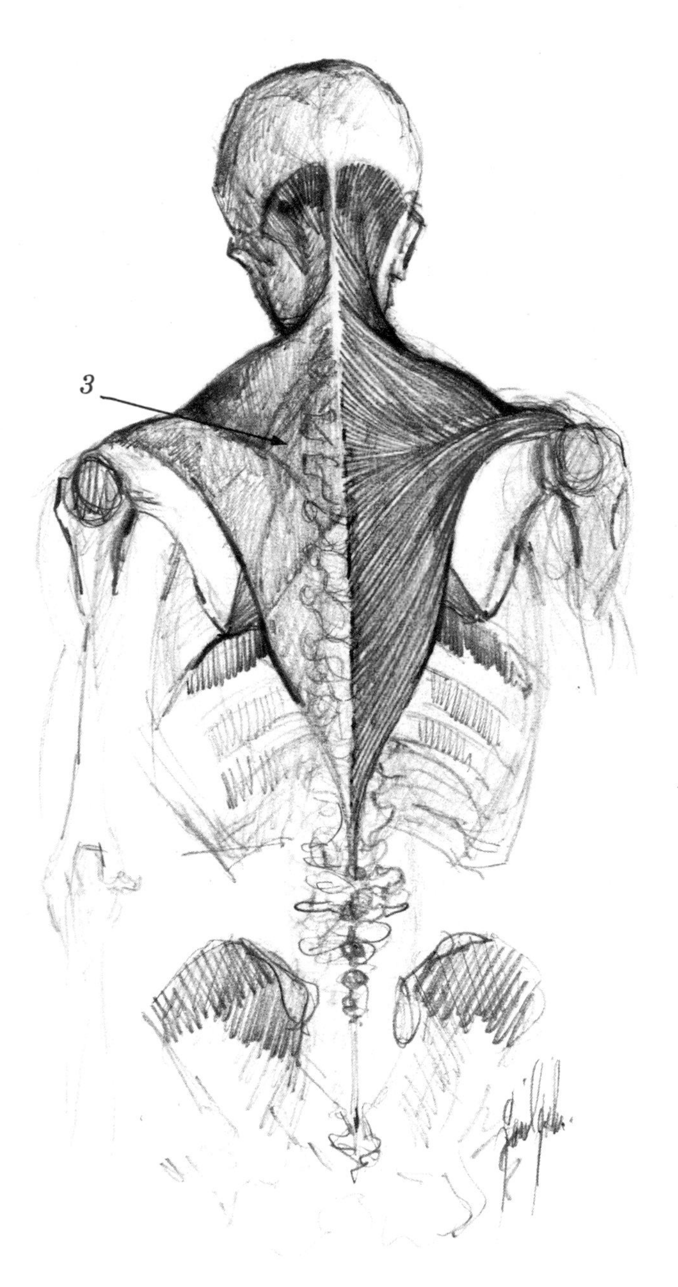

Fig. 200.

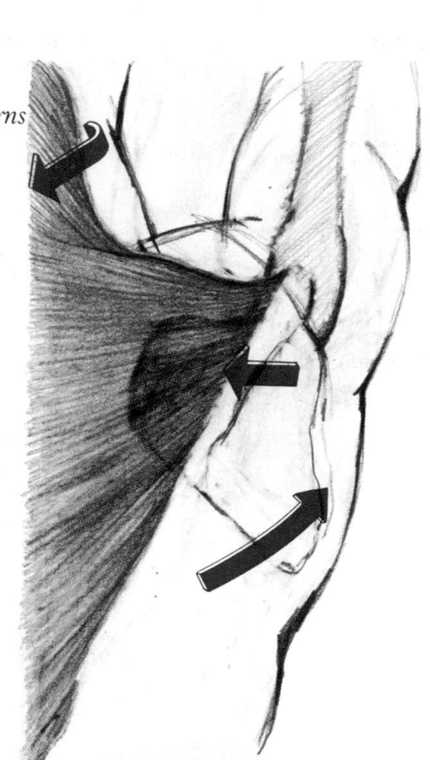

Fig. 201.

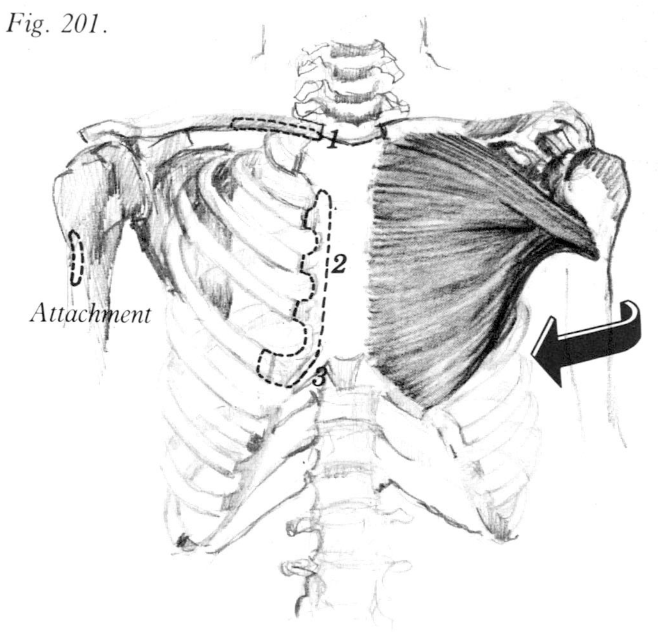

Group C

The large, flat, muscles which arise from the trunk and pass to the upper arm. They are the most important muscles as far as strength and flexibility are concerned.

1. Greater chest muscle (Pectoralis major; pectoral = chest)

Origin (Figure 201): (1) Inner part of the collar bone, (2) manubrium and sternum and (3) part of the costal cartilage (of the ribs).

Insertion: External area of attachment of the upper arm.

Function: Forms the anterior wall of the armpit, and adducts the arm and rotates it outwards. It pulls a raised arm down and swings a lowered arm forward (flexion of the shoulder).

Exercises for the greater chest muscle (pectoralis major) are shown below.

Fig. 202.

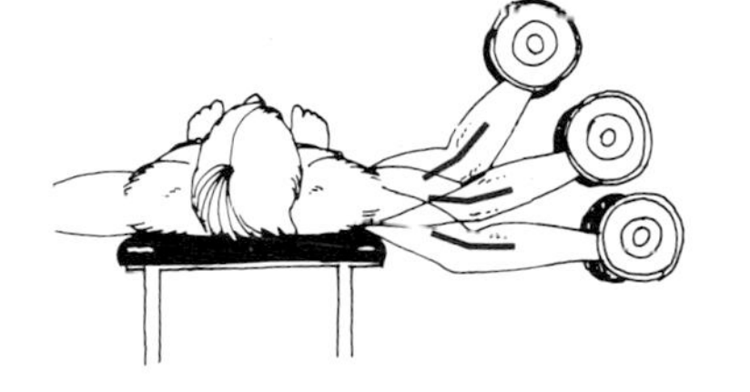

Lie on your back on a bench with dumbbells in your hands (Figure 202). Lift the dumbbells and keep your arms slightly bent in order to avoid overexerting your elbows. The upward movement provides concentric training, and the downward movement provides eccentric training. You can vary the work load by varying the extent to which you bend your elbows. By keeping your arms a little straighter on the way down, you subject the muscle to a little more stress.

Fig. 203.

The same exercise as in Figure 202. This exercise trains only the upper part of the muscle (1) namely, that part which has its origin on the collar bone (i.e. prevents the arm from falling downwards). Other muscles taking part in the exercise are the deltoid muscle (Figure 207) and biceps brachii (see below) (Figure 217).

Fig. 204.

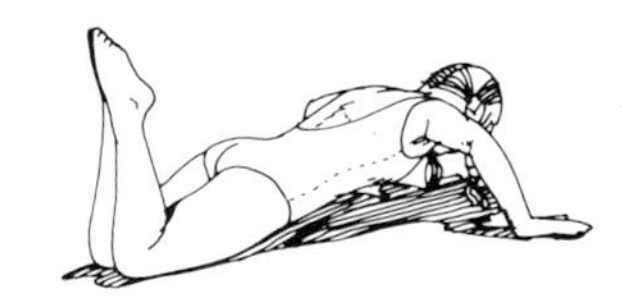

Bending and stretching the arms while the hands are well separated, activates pectoralis major (and triceps brachii, deltoid, serratus anterior and trapezius). Static training is provided when the palms of the hands are pressed together. The position of the hands in front of the body should be varied.

Fig. 205. *Fig. 206.*

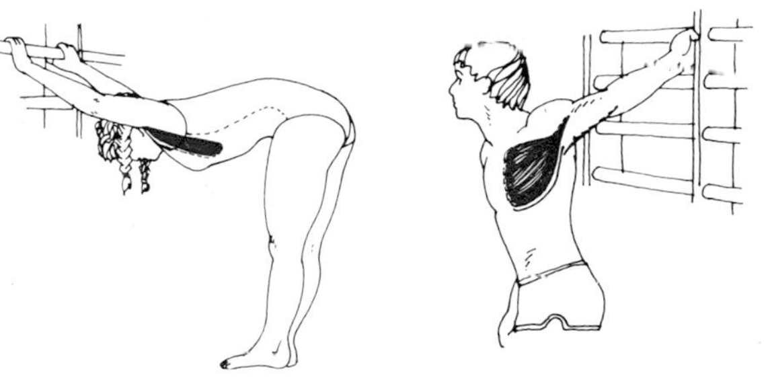

A good way to round off these exercises is to stretch the greater chest muscle following the PNF-method (Figures 205 and 206).

2. Deltoid muscle

Fig. 207.
(a)

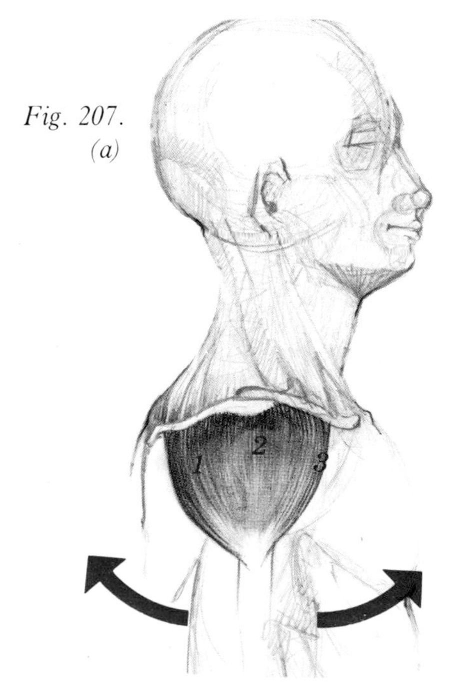

Fig. 207.
(b)

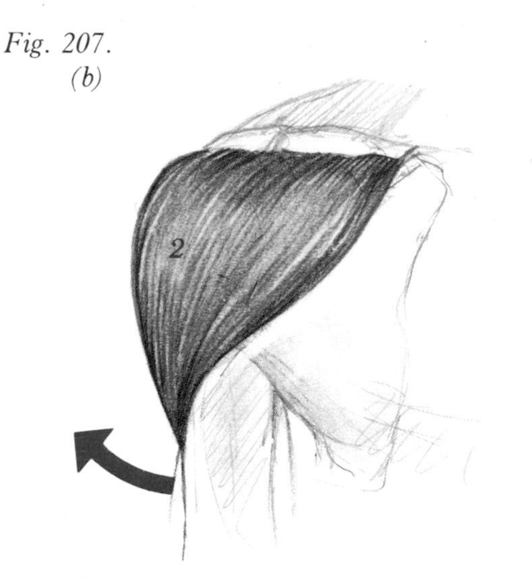

Origin: Outer part of the collar bone, and along the entire spine of the posterior surface of the shoulder blade. (May even be assigned to group A.)

Insertion: Along the shaft of the upper arm (deltoid tuberosity).

Function: Since its origin arches around the shoulder joint, this muscle can take part in all the movements of the arms. Its most important function is to (2) lift the arm straight outwards and upwards (abduction). The parts of the muscle which arise from the posterior surface of the shoulder blade (1) swing the arms backwards and rotate it outwards. The parts which arise from the collar bone (3) swing the arm forward and rotate it inwards.

Fig. 208.

The deltoid is trained in almost all exercises which involve the arm. By lifting as shown in Figure 208, you can train the entire muscle. The work load should not exceed your ability to lift — with almost straight arms — the full distance six to eight times. Lift and lower your arms at the same pace (supraspinatus is also trained in this exercise, see Figure 194).

3. Broad back muscle (latissimus dorsi)

Fig. 209.

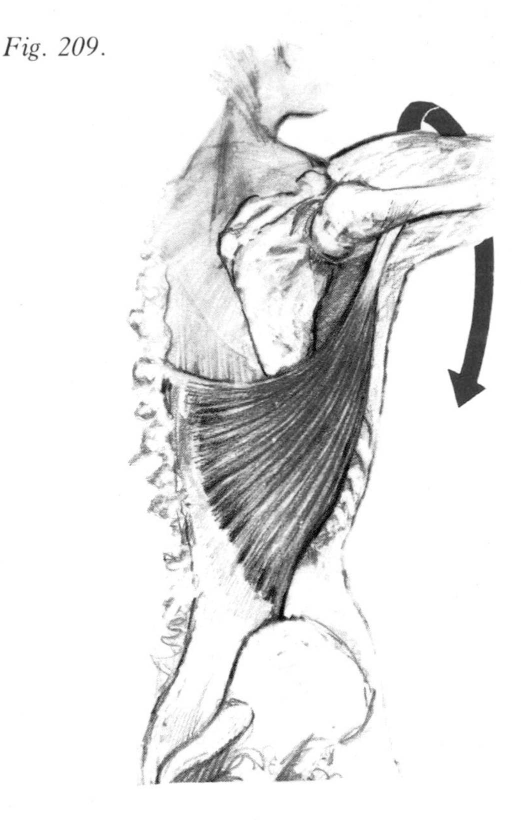

Origin: The spine of the lower half of the spinal column down to the sacrum and from it out to the crest of the hip bone.

Insertion: Anterior attachment area of the upper arm.

Function: Forms the back wall of the armpit and pulls the arm in behind the back (i.e. swings it backwards and rotates it inwards).

There is usually equipment for training the latissimus in the majority of bodybuilding gymnasiums. Using the special latissimus training apparatus, you should be able to cope with a work load of about six to eight exercises. (For maximal strength.)

By pulling yourself up so that your head comes in front of the bar [Figure 210 (b)], you will train exactly the same muscles as in (a), but with a greater work load (total bodyweight). The biceps brachii (arm bending), trapezius (shoulder blade positioning) and the pectoralis major (upper arm adduction) also take part in this exercise.

Fig. 210.

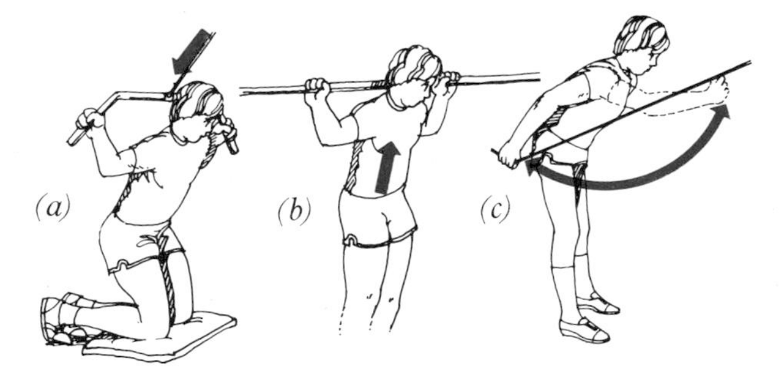

Figure 210 (c) shows how you can train the lattisimus dorsi by pulling on a suspended weight with a rubber hose. In general, this muscle does not need flexibility training. If you bend forcefully to one side and hold your arms outstretched above your head, you may possibly feel the latissimus blocking the movement.

The exercises below are examples of pure shoulder rotation.

Lie on your stomach on a bench with your elbows pointing outwards, and lift a dumbbell backwards (inward rotation) and forwards (outward rotation).

Fig. 211.

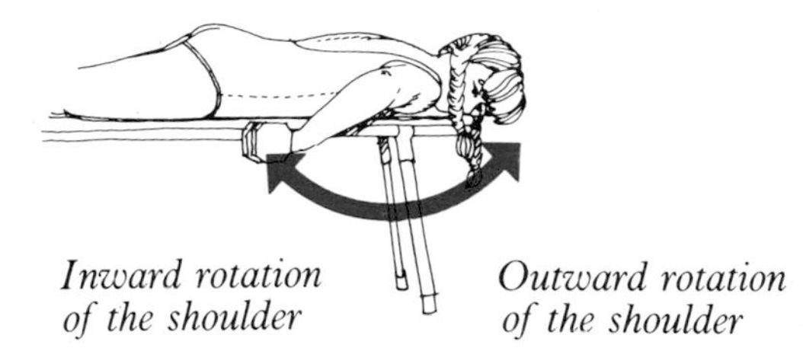

The main muscles helping to inwardly rotate are:

subscapularis,
pectoralis major,
latissimus dorsi,
teres major.

The main outward rotators are:

infraspinatus,
teres minor.

Flexibility of upper arm inward and outward rotation can be tested in the position presented in Fig. 212.

Fig. 212.

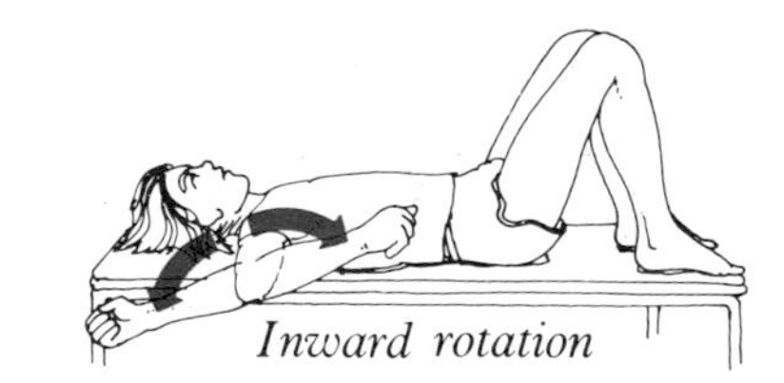

Anterior serrated muscle (serratus anterior)

This plays a very important role in stabilising the shoulder. It arises from eight, nine or even ten ribs, and passes backwards along the rib cage in behind the shoulder blade to be inserted into the medial border of the shoulder blade (should be assigned to group B).

Serratus anterior prevents the shoulder blade from being pressed backwards when the body supports itself on its arms [Figure 213 (a)]. It is trained in all types of exercises where the body is supported by the arms [Figure 213 (c)].

Fig. 213.

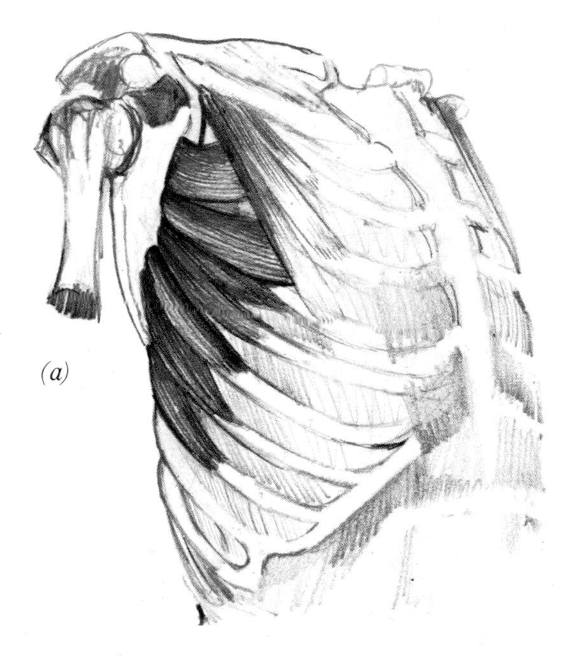

Figure 213 (b) shows how the shoulder blade moves when the serrated muscle contracts.

Fig. 213.
(b)

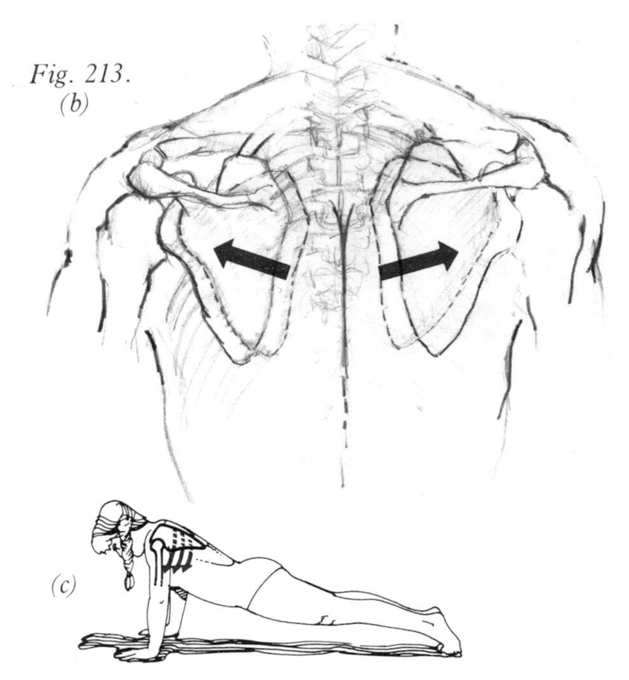

(c)

The two muscles shown in Figure 214 play no significant role in athletics but nevertheless contribute to stabilising the shoulder blade.

Fig. 214.

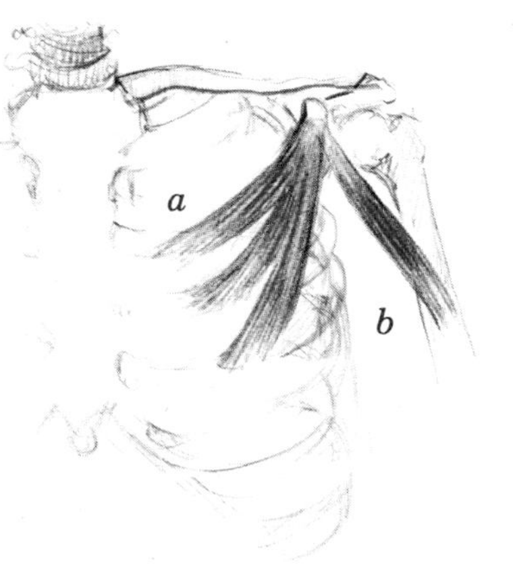

Figure 214 presents the location of (a) pectoralis minor (lesser breast muscle) and (b) coracobrachialis.

Elbow Joint

Elbow movements occur at three separate joints, namely:

hinge joint between the upper arm and ulna,
pivot joint between the ulna and radius and
ball and socket joint between the upper arm and radius.

Fig. 215.

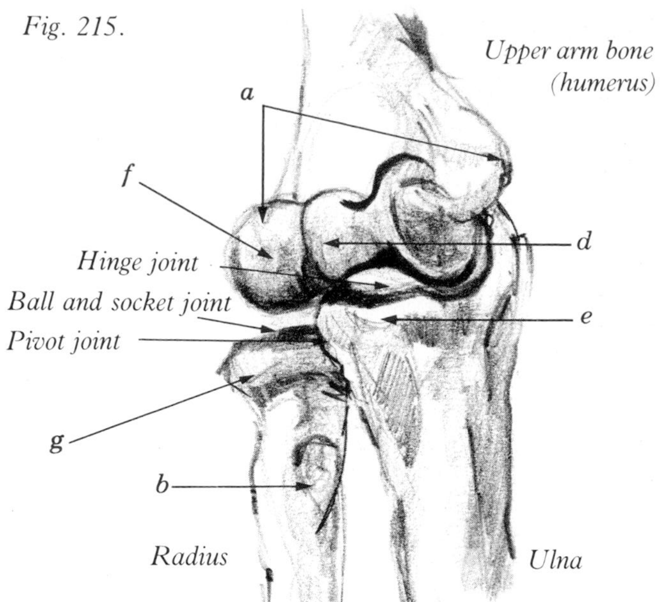

Right elbow viewed from the front (with the palm of the hand facing forward).

Fig. 216.

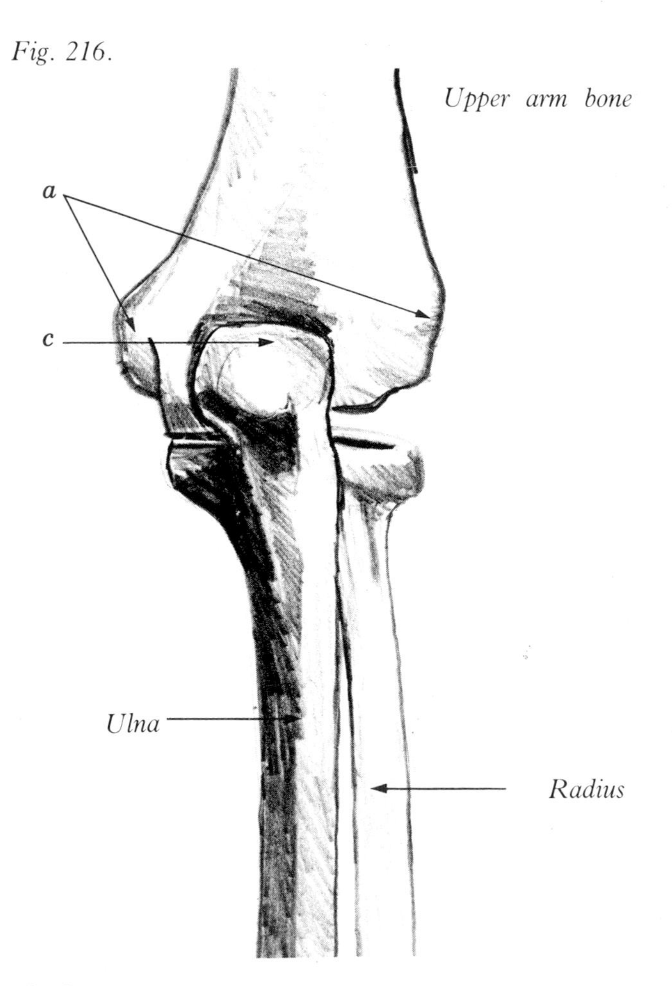

Right elbow viewed from the back

The projections and articular surfaces concerned with the function of the elbow joint can be seen in Figures 215 and 216.

(a) Epicondyles give attachment to the muscles which move the hand at the wrist joint.
(b) Radial tuberosity gives attachment to the biceps brachii [Figure 217 (a)].
(c) Elbow outgrowth (olecranon process) gives attachment to triceps brachii (p.89).
(d) The articular surface of the upper arm bone at its hinge joint with the elbow (trochlea = pulley).
(e) Articular surface of the ulna at its hinge joint with the elbow (coronoid process = crown projection).
(f) Articular surface of the ball and socket joint (capitulum = small head).
(g) Head of the radius which forms the "socket" part of the ball and socket joint.

A relatively common difference between men and women can be seen in the elbow projection (c) and the hollow of the upper arm bone into which it is received when the arm is stretched. Because this projection and its corresponding hollow differ, women are more prone to overstretch their elbow joint than men.

Elbow flexion

The three most important flexors are the following.
(a) Two-headed arm muscle (biceps brachii)
(Bi = two; ceps = head; brachii = upper arm.)

Origin:
1. Caracoid process ("short" tendon).
2. Upper rim of the shoulder blade's articular surface ("long" tendon).
Insertion:
3. Radial tuberosity (a roughened surface at the upper end of the shaft).
Function:
A. Flexes the elbow.
B. Turns the forearm so that the palm of the hand points upward (supination).
C. Swings the upper arm forward (flexion of the shoulder joint).

Fig. 217.

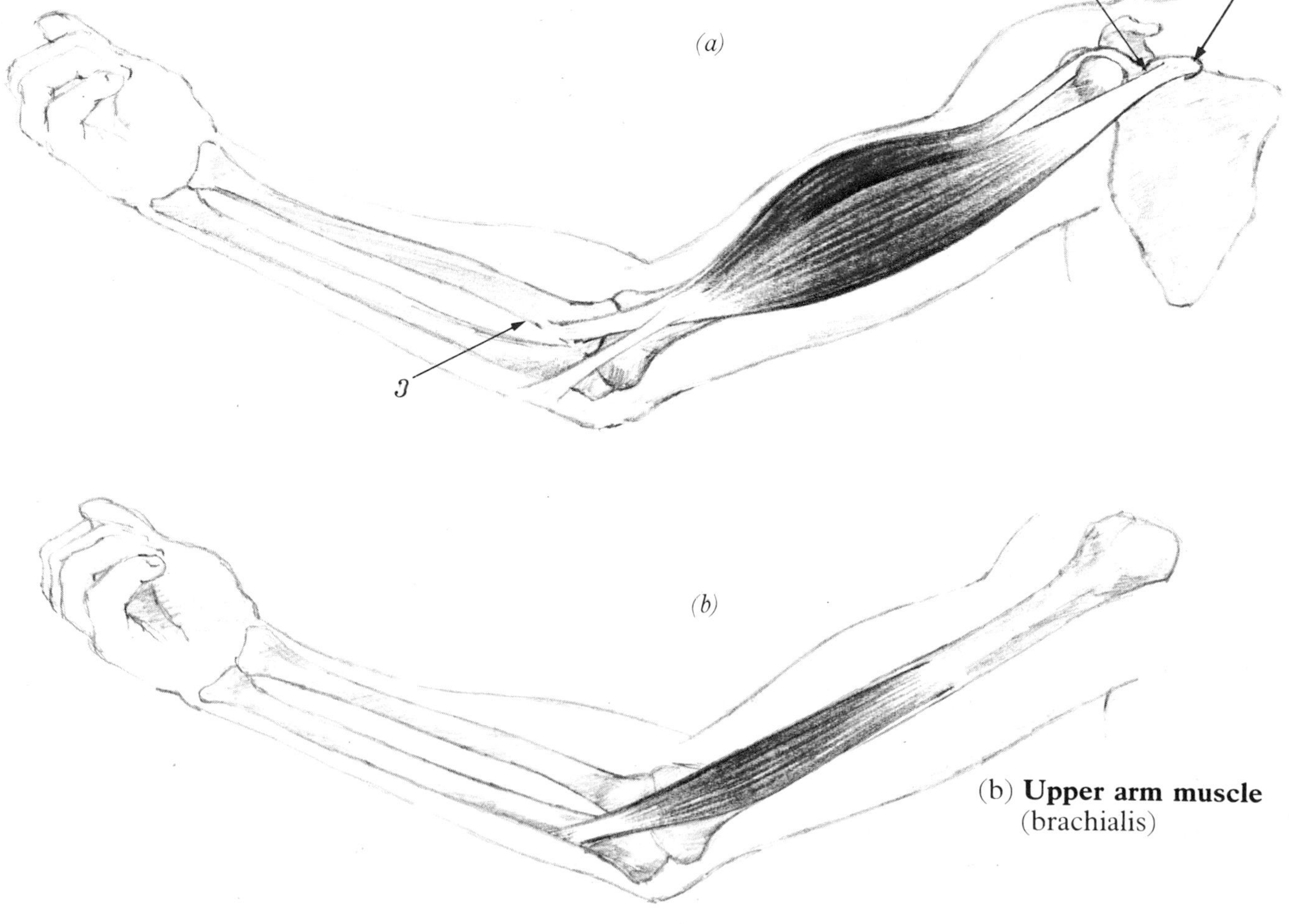

(b) **Upper arm muscle** (brachialis)

Fig. 217.

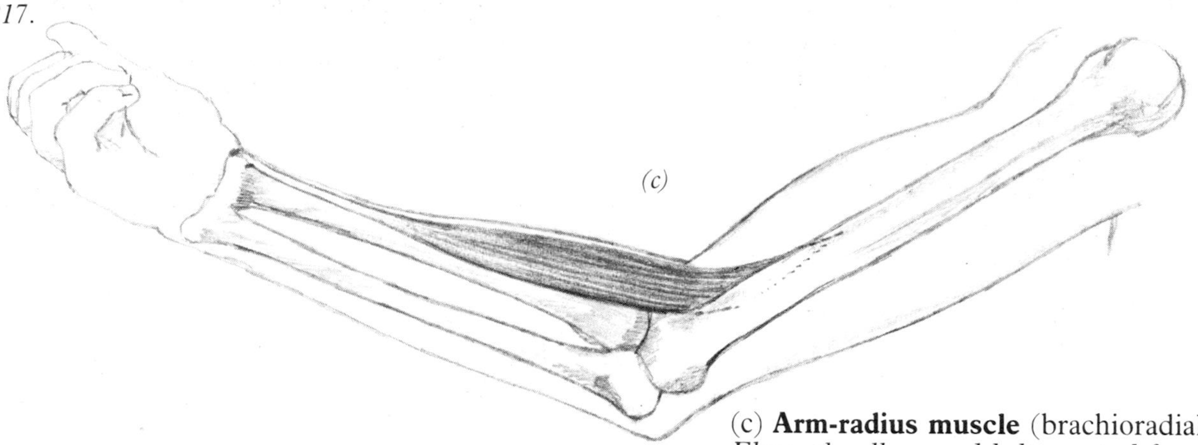

(c)

(c) **Arm-radius muscle** (brachioradialis) *Flexes the elbow and helps control forearm rotation.*

Fig. 218.

In order to develop strength in the elbow flexors, the barbell (or whatever) must be grasped with the palms of the hands turned upwards, since this is the only way the biceps can work with maximal strength.

In Figure 218 the girl's hands are supinated (they grasp the bar from underneath). Therefore all her flexors work maximally.

Fig. 219.

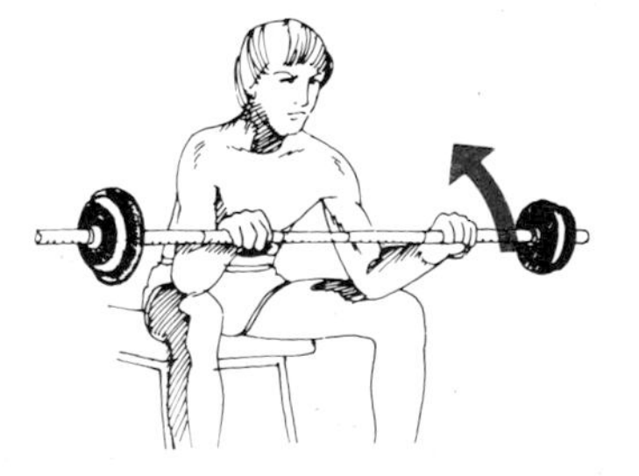

In Figure 219 the boy's hands are pronated (they grasp the bar from above), which means that his biceps cannot work maximally. Instead, he subjects his other flexors to a heavier work load. In general training, one should alternate between these two methods. Figure 220 shows arm pull-ups with (a) pronated forearm and (b) supinated forearm.

Fig. 220.

(a)

(b)

The long tendon of origin of the biceps [marked "2" in Figures 217 (a) and 221] passes through the shoulder joint and emerges from the joint capsule at its attachment to the upper arm bone where it descends in a groove between the anterior projection (lesser tuberosity) and the outer projection (greater tuberosity) (Figure 193). Because of this, the biceps contributes greatly to the stability of the shoulder joint. A person who has dislocated his shoulder has, therefore, good reason to build up the strength of his biceps muscle since by doing so he may help to prevent recurrent dislocations.

Biceps strength depends, amongst other things, on the position of the upper arm. If we consider the distance between the origin and insertion of the biceps muscle, we see that it is greater when the arm points straight upwards than when it hangs down.

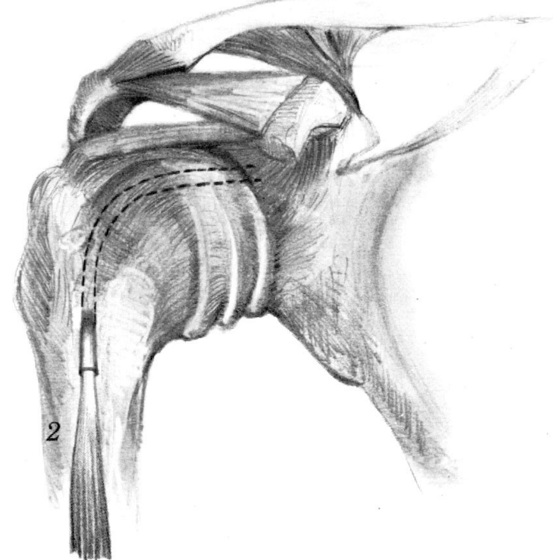

Fig. 221.

Thus, the capacity of the biceps to bend the arm with force is greater when the arm is directed upwards. The numbers in Figure 222 give the approximate relationship between strength to carry, pull towards the body and lift oneself up.

A reasonable conclusion we can draw from Figure 222 is that the human body is better suited to hang in trees than to carry buckets.

Elbow extension

The elbow extensor (triceps brachii) is attached to the elbow outgrowth (olecranon). It has three heads. One arises from the shoulder blade and the other two from the posterior surface of the upper arm.

Its function is to straighten or extend the elbow and to swing the arm backwards.

A person's strength differs in different positions depending both on the length of his muscles (p.18) and the size of his moment arm into the elbow joint. Notice that the muscle is attached far back on the elbow outgrowth in order to allow the most favourable moment arm possible in all elbow joint positions.

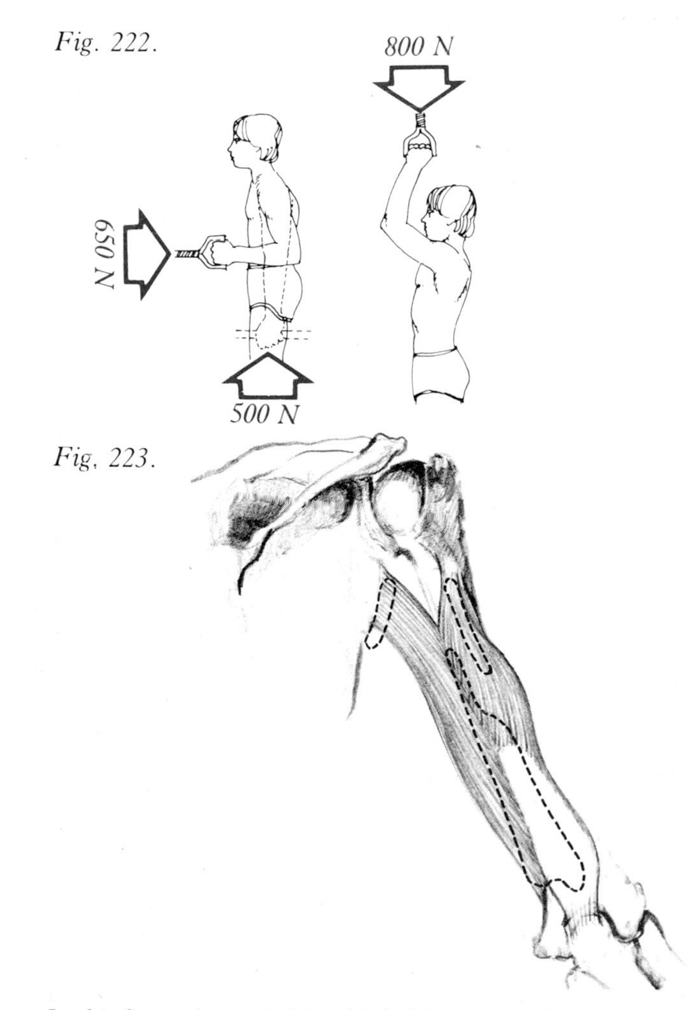

Fig. 222.

Fig. 223.

In this figure the medial head is hidden behind the long head.

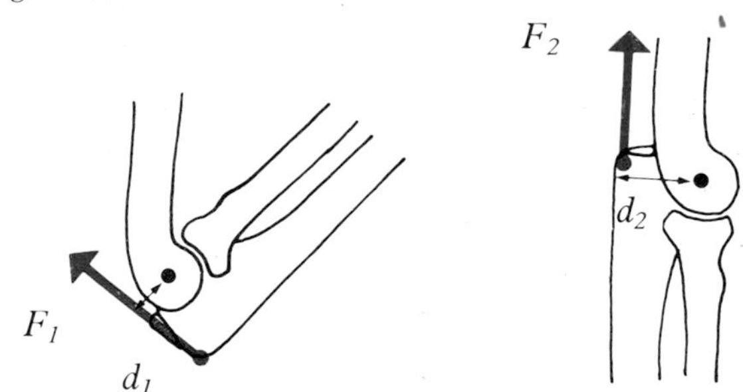

Fig. 224.

The extensor's lever arm is least (d_1) when the arm is fully flexed. At the same time, the force of the muscle is greatest. Strength (moment of force) is least when the joint is extended because the muscle force (F_2) is small. The strength of the muscle also depends on the way the arm is held in relation to the shoulder blade. If the arm is stretched out in front of the body, the long head of the triceps is in a worse position (shorter distance between origin and insertion) than if the arm is held straight above the body. Figure 225 shows the approximate relationship between the strength of the triceps brachii in downward, forward, and upward presses.

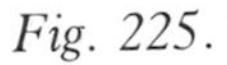

Fig. 225.

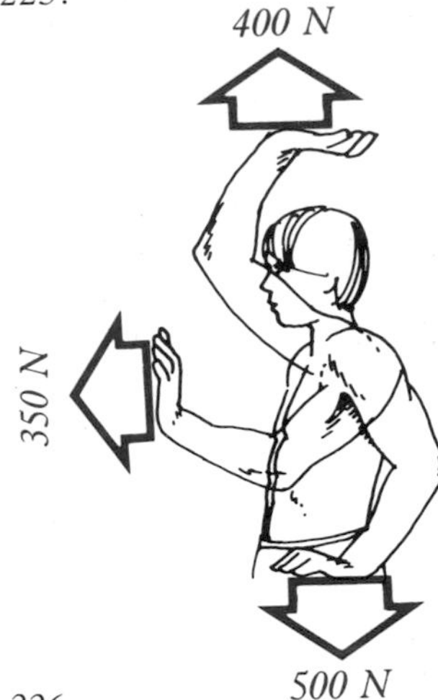

Fig. 226.

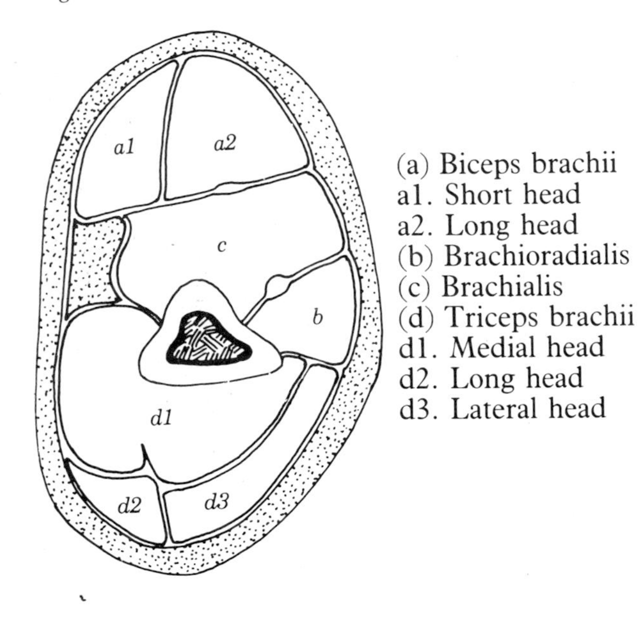

Figure 226 is a cross section to show the location and size of the muscles of the right upper arm.

Figure 227 gives suggestions for different exercises for developing strength in the triceps.

Fig. 227.

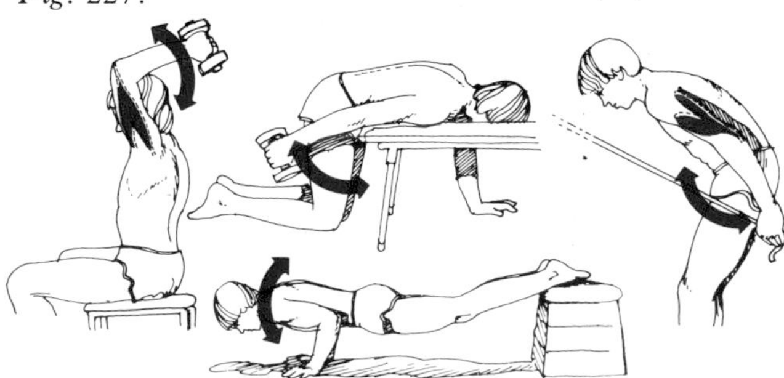

Wrist joint

The skeleton of the hand consists of eight carpal bones (bones of the wrist), five metacarpal bones (bones of the hand) and 14 phalanges (bones of the fingers).

The hand changes position by movement of the wrist which is a condyloid joint. It is formed by the articulation of three of the wrist bones with the radius.

Fig. 228.

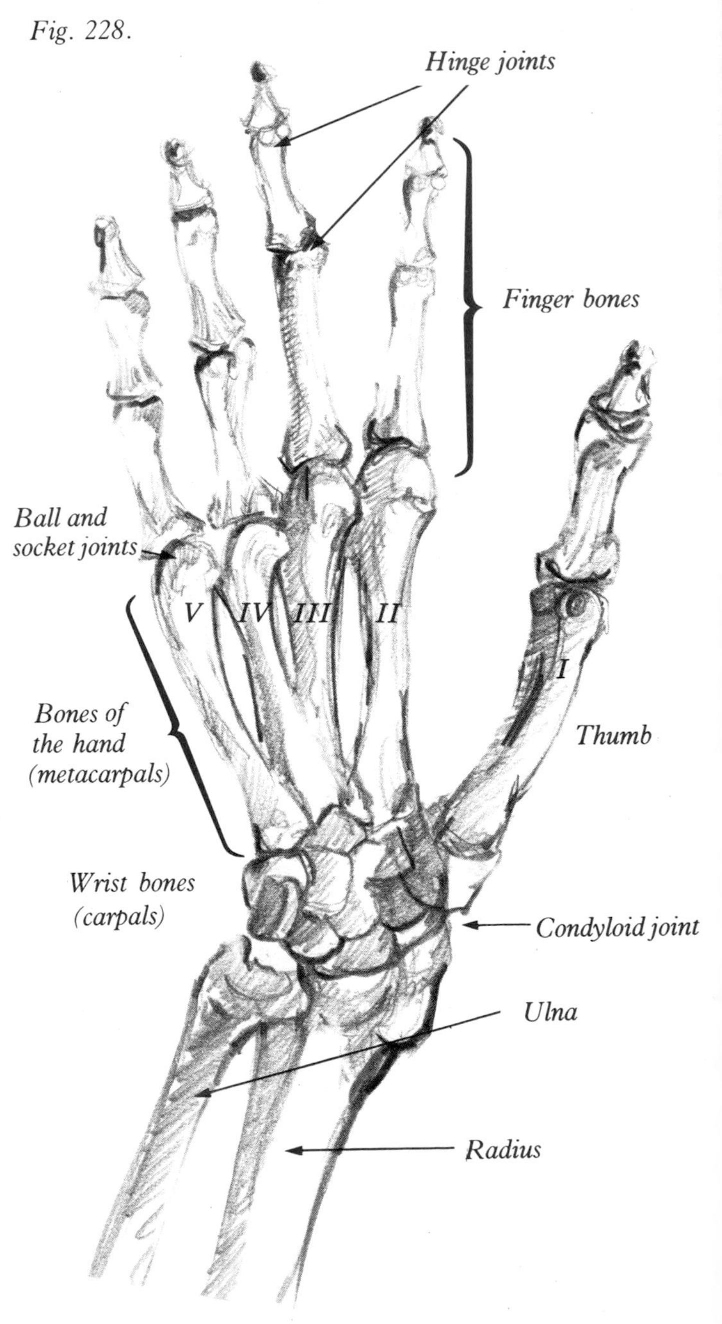

Right hand with palm turned forward [supinated according to Fig. 229 (f)].

The movements allowed at the wrist joint are (Figure 229):

(a) flexion (palms towards you),
(b) extension (palms away from you),
(c) abduction (flat hand tilted outwards) and
(d) adduction (flat hand tilted inwards).

Fig. 229.

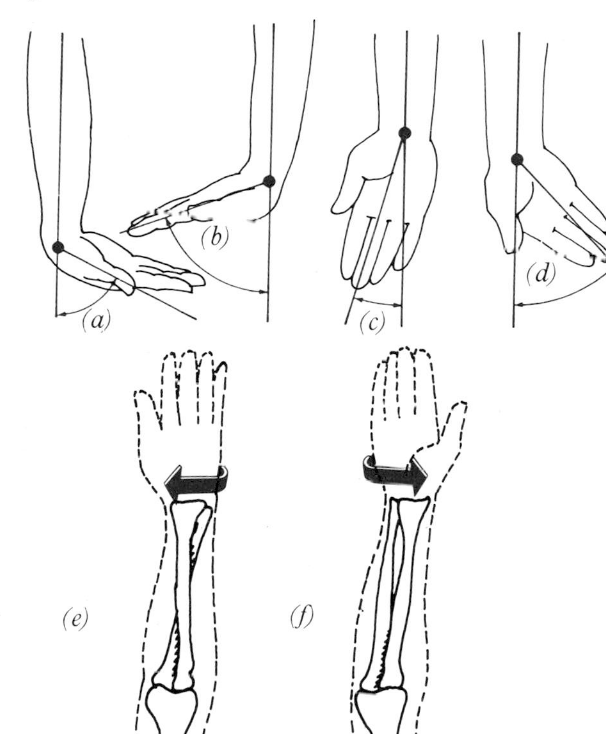

The movements involved in turning the hand are called (Figure 229):

(e) pronation and
(f) supination.

Supination and pronation occur between the bones of the forearm (at the pivot joint formed by the ulna and radius, see Figure 215), and not at the wrist joint. The muscles which bring about rotation of the forearm, and thereby turn the hand, are called pronators and supinators. The diagrams below present the most important of these.

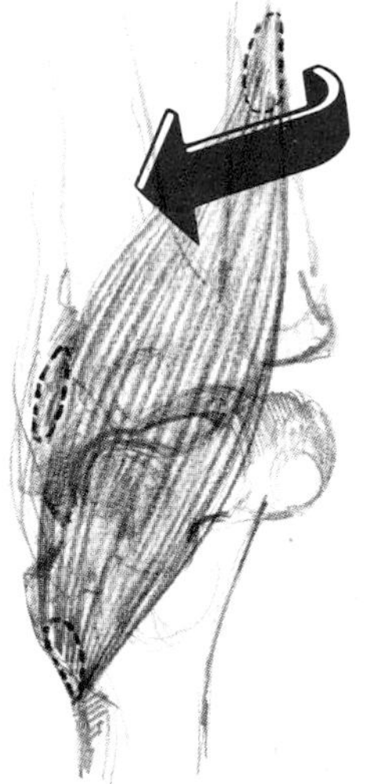

Fig. 231.
Pronator teres. Right elbow viewed as in Fig. 229 (f)

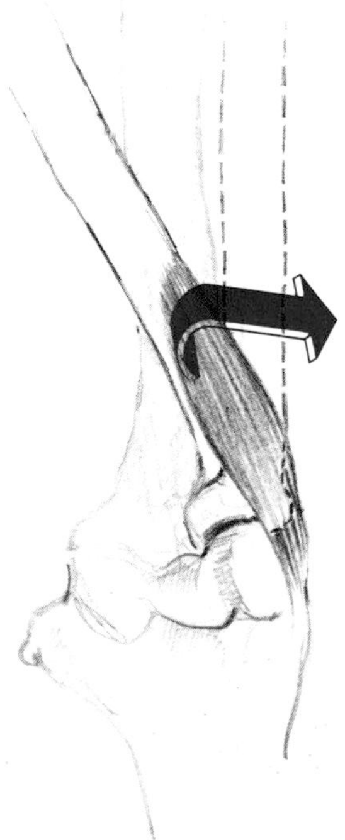

Fig. 230.

Supinator
Right elbow viewed as in Fig. 229 (e)

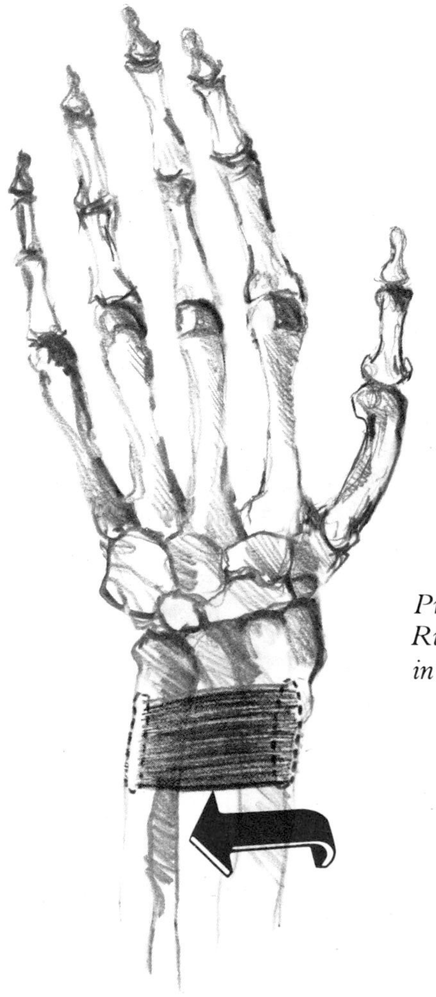

Fig. 232.

Pronator quadratus
Right hand viewed as in Fig. 229 (f)

However, by far the strongest supinator of the hand is the biceps brachii (p.87).

The supinators are stronger (have a greater moment of force) than the pronators. This relationship has been taken into account in the manufacture of such things as screws. Thus, the thread is constructed in such a way that a right-handed person uses supinated movements to screw it into an object, and thereby allows his muscles to work concentrically. He can easy feel how the biceps brachii works with great force. If the screw is so troublesome that he cannot quite manage it, he locks his wrist (lets supinators work statically (or isometrically) — see the curve on p.20), and thus produces rotation by bending his elbow and rotating his shoulder joint outwards.

Of the 20 or so muscles responsible for the wrist's movements, the largest and most important are those attached to the epicondyles of the upper arm bone (humerus). The abductórs and extensors of the wrist arise from the external condyle.

Fig. 233.

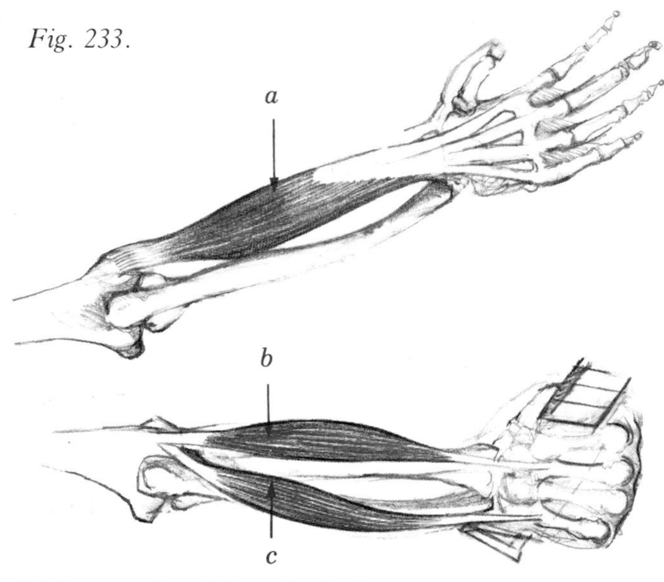

Right arm viewed from the back

Figure 233 shows some muscles that are brought into action when a person hits a backhand or lifts a hammer. The concept "tennis elbow" actually means that the area of origin of these muscles is injured.

The following can be seen in Figure 233:

(a) finger extensor (extensor digitorum),
(b) radial-wrist extensor (extensor carpi radialis); passes down the radius and
(c) ulna-wrist extensor (extensor carpi ulnaris); passes down the ulna.

The adductors and flexors of the wrist arise from the internal epicondyle. Adduction and flexion are the movements responsible for such actions as throwing or hitting a forehand smash. Injuries caused by overexertion in these types of activities are called thrower's elbow, golfer's elbow, etc.

All the muscles in Figures 234 and 235 arise from the internal epicondyle:

(d) finger flexor (flexor digitorum),
(e) radial-wrist flexor (flexor carpi radialis) and
(f) ulna-wrist flexor (flexor carpi ulnaris).

Fig. 234.

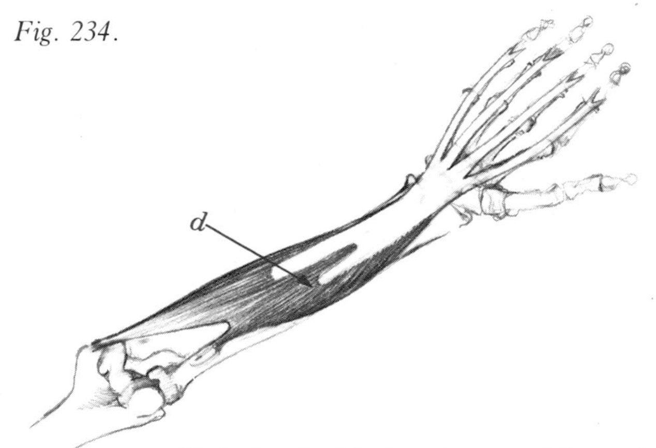

Right hand with the palm turned forward

For the prevention of such injuries as tennis elbow, strength should be trained in moderate doses. The exercises should subject the elbow to only light work loads. They should provide both static and dynamic training. Movements should be produced in all directions allowed by the wrist. Examples of suitable loads for these exercises are dumbbells, clubs and rackets.

Fig. 235.

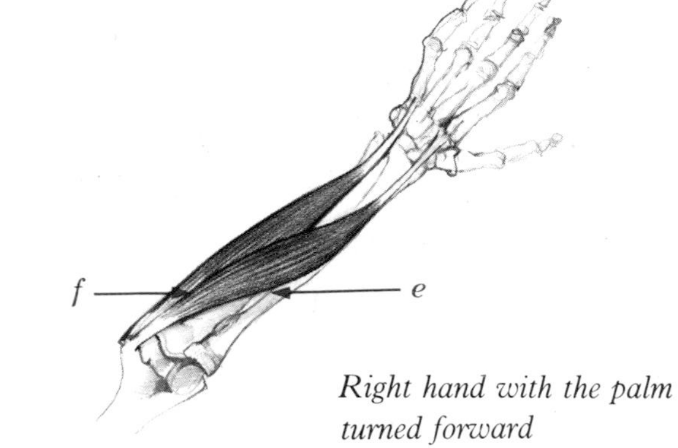

Right hand with the palm turned forward

Fig. 236.

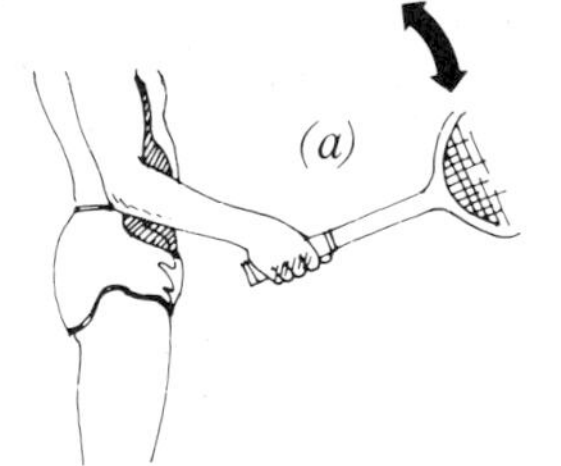

Towards the thumb (abduction)

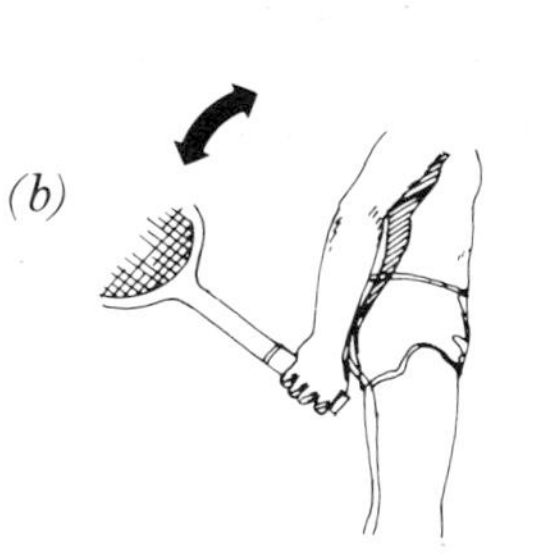

Towards the little finger (adduction)

Bending (flexion)

Stretching (extension)

Fig. 237.

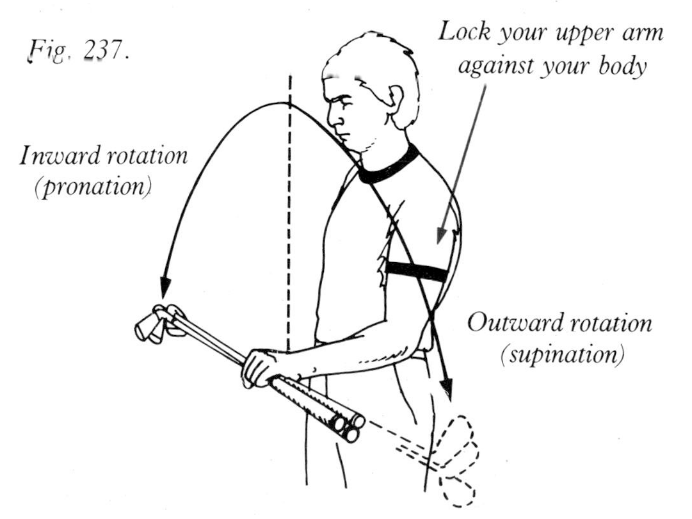

The following movements can be recommended for limbering up the wrists and possibly stretching the wrist muscles.

Fig. 238.

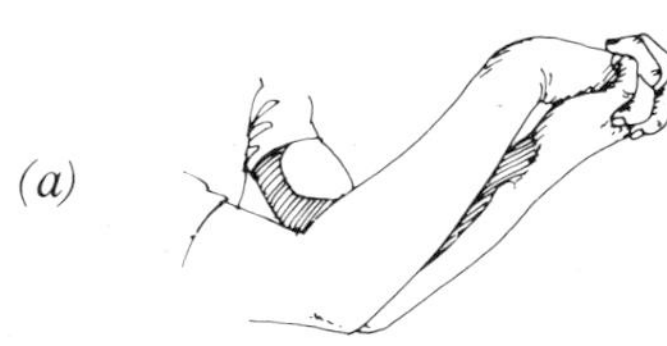

(a) Lock your hands loosely and roll them around.

(b) Place your hands on the ground and press lightly. Keep your arms straight. Try to hold your wrists at right angles to your arms.

(c) As in (b), except that you could press the back of your hand against a wall, or use your other hand as shown in the diagram.

Notice that movements involving flexion (c) are somewhat less flexible than those involving extension (b).

Chapter 6

SPORTS MECHANICS

A basic understanding of the laws of mechanics is necessary for the study of movement. When we are familiar with these laws, we can more readily analyse and design suitable exercises. Unfortunately, when athletic performance is described with the aid of the laws of mechanics, the discussion tends to be too theoretical. In this section we will consider a few concepts that are directly applicable to the practical understanding of how and why exercises are performed in a certain way. The concepts I, force; II, centre of gravity; III, moment of inertia, will be described with the aid of diagrams (not formulae). The concept "moment of force" has already been defined on pp.30–35. In addition, it was used in a number of examples in the sections on anatomy (pp.51, 66, 67 and 75). Therefore, moment of force will only be illustrated in this chapter by providing some further examples.

I. Force

Fig. 239.

A force is represented by an arrow (vector). The magnitude of the force is indicated by the length of the arrow shaft and the direction of the force is indicated by the arrow head.

A distinction is made between internal and external forces. The external forces which must be taken into account when studying sports are (Figure 239):

(a) gravitational pull on a body *(mg)*,
(b) force exerted on a body by the ground (called normal force, *N*),
(c) friction between feet and ground (F_μ) and
(d) air resistance (F_a).

Internal forces are:

(e) muscle forces (F_m) and
(f) tendon, ligament and connective tissue forces *(F)*.

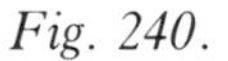

Fig. 240.

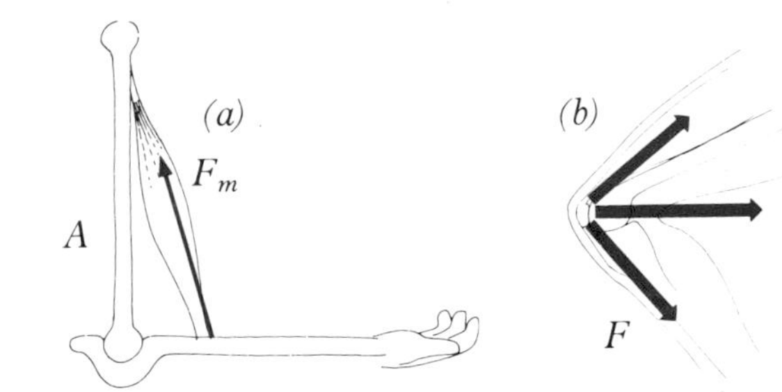

We will now look at some of the properties of these different types of forces.

External Forces

(a) Force of gravity

Fig. 241.

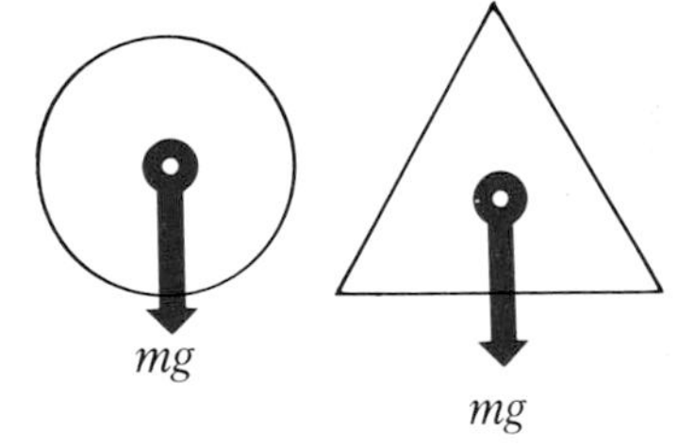

The force of gravity can be thought of as acting on a body at the exact centre of its mass (p.101). This force results from the action of the earth drawing all objects towards its centre. The size of the force depends on the weight or mass of the body being acted upon. Sixty kilogrammes corresponds to about 600 N. Kilogrammes are converted into newtons by using the following relationship: 1 kg = 9.81 N. This conversion factor is represented by the letter *g*. The gravitational force acting on a body weighing *m* kilogrammes is thus, *m* x *g* newtons, and is written *mg*.

Fig. 242.

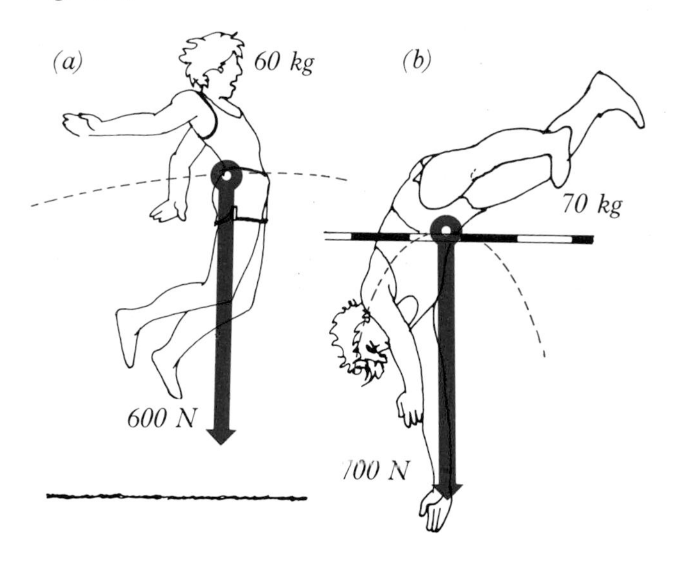

(b) Normal force (*N*)

Fig. 243.

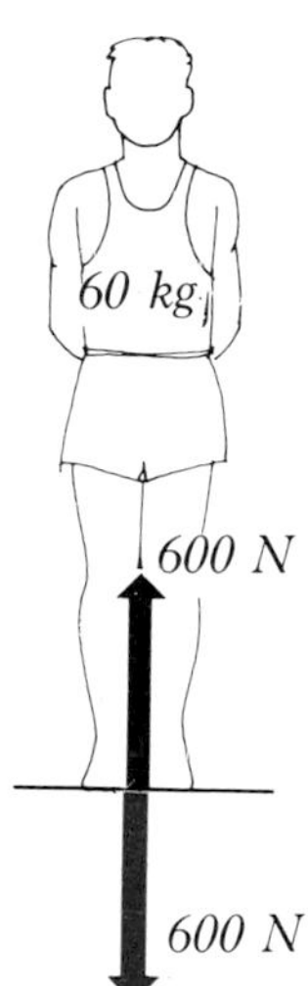

Normal force is exerted on a body when it comes in contact with the ground. If a person weighing 60 kg, stands still on flat ground, he will be exerting a force of 600 N on the ground. The reaction force exerted on his body by the ground will also be 600 N. Thus, these two forces are equal in size but act in opposite directions and on different objects. They constitute a so-called pair of co-linear forces, where one of the forces is called "action" and the other the "reaction". Some further examples of action – reaction forces are given on p.97.

The magnitude of the normal force depends on the force with which a person presses on the ground (i.e. how heavy he is) and to what extent he activates his hip, knee and ankle extensors. In taking off for a jump, swinging a leg forward while running and landing from a jump, the normal force can be as much as three to four times the body's own weight (*mg*).

Fig. 244.

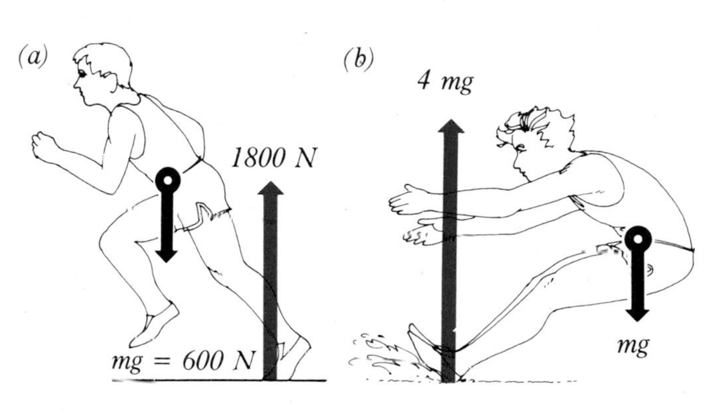

If a person stands and bounces up and down, the normal force acting on him will change. This can be explained by analysing how muscles work. To the skier bending and stretching his legs in Figure 245 the following applies.

His muscles work statically as long as he skis without bending. The normal force is then 600 N if he weighs 60 kg. From 1 to 2 he relaxes his muscles a little, with the result that less pressure is exerted on the ground (i.e. the normal force is reduced to, say, 400 N). From 2 to 3 he halts the downward movement which was present at 2. He accomplishes this by working his muscles eccentrically and

Fig. 245.

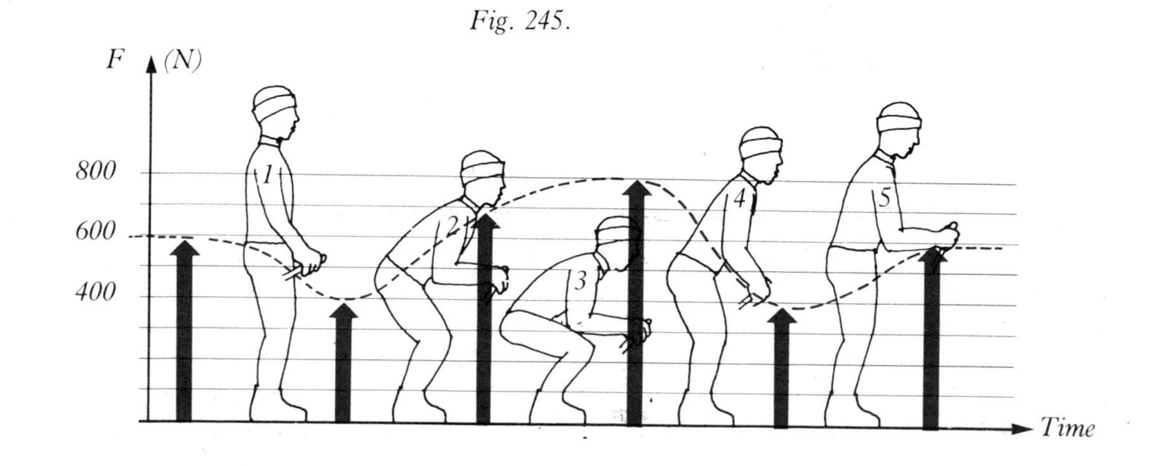

pressing more forcefully on the ground. From 3 to 4 his muscles work concentrically, and they still press forcefully on the ground (max. 800 N according to our example). At 4 the skier has such upward acceleration that he reaches 5 without pressing especially hard. He can even enjoy a moment's relaxation (400 N). From 5 onwards, he stands still again and his work load will once again be 600 N because his muscles work statically.

Fig. 247.

Fig. 246.

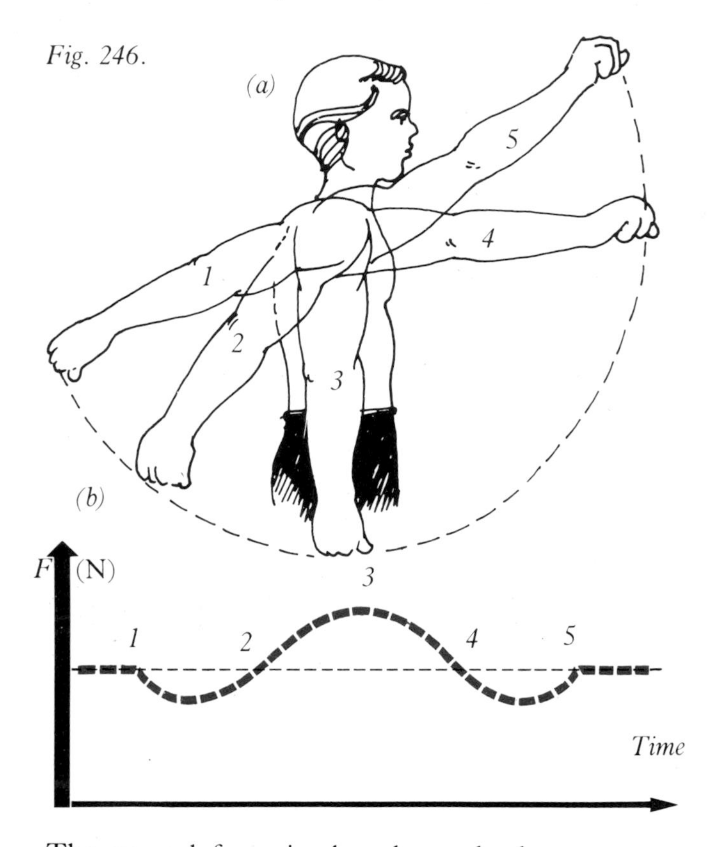

The normal force is also changed when a person swings his arms (Figure 246). Arm movements serve to load and unload the body during different activities. What is the significance of "arm swinging" in running, jumping, landing from a jump, clearing a bar, downhill skiing, kicking a football, vaulting, balancing, etc.? Examine an activity that you are familiar with, and think about the function of the arm movements (Figure 247).

In general, it can be said that a force is needed to change the velocity of the arm. The origin and insertion of the arm muscles are then acted upon by equal but opposing forces (i.e. the arm and the body are affected in different directions see Figure 248).

When the arm is accelerated or halted by, say, the greater chest muscle (pectoralis major) or the broad back muscle (latissimus dorsi), the body is affected in the opposite direction. The arrows in Figure 249 show how the body is affected when the velocity of the arm changes. The black arrow represents the force acting on the arm and the red arrow represents the force acting on the trunk. The magnitude of the force is the same at the origin of a muscle as it is at its insertion.

Fig. 248.

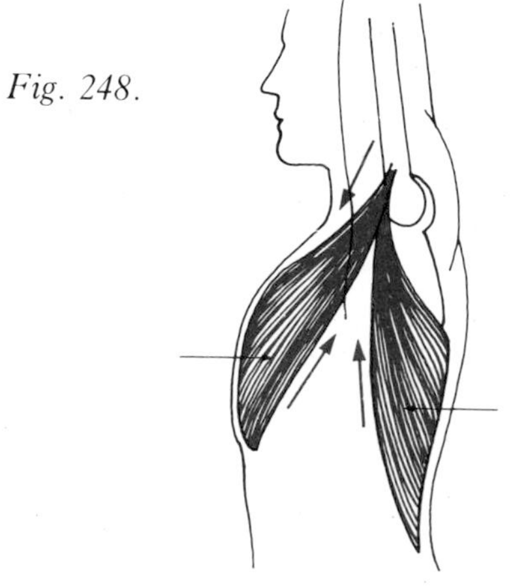

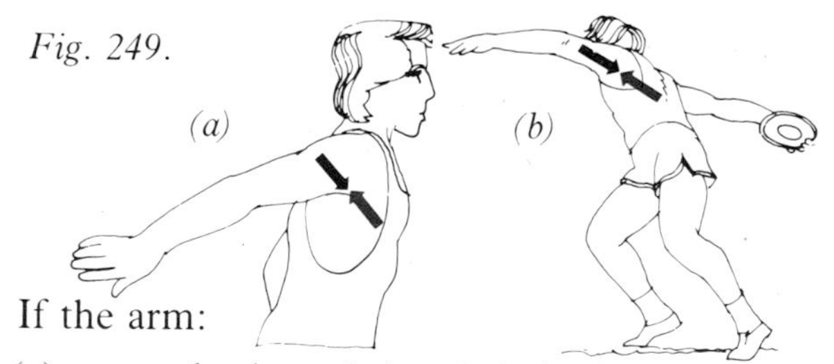

Fig. 249.

If the arm:

(a) moves backwards but is halted,
(b) accelerates backwards,

(c) accelerates forwards or
(d) moves forwards but is halted,
then the body will be affected as shown by the red arrows.

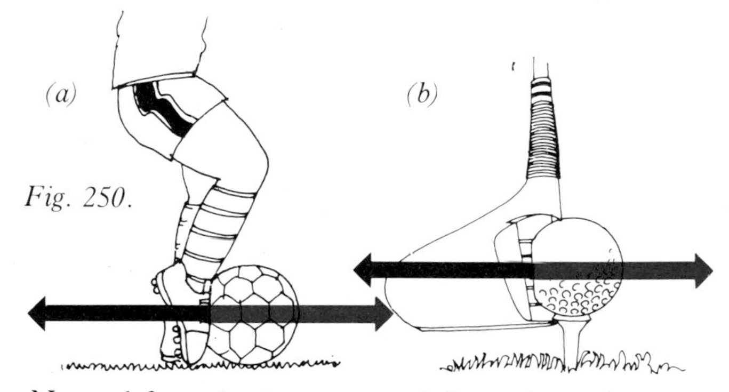

Fig. 250.

Normal force is also a part of the action of striking (a football, golf ball, tennis ball). The normal force acting on the ball is of the same magnitude as the normal force acting on the foot [Figure 250 (a)]. It is said that these normal forces are each others' action–reaction forces. The force acting on the ball increases to a maximum when the ball is maximally compressed, and thereafter decreases to zero. The situation is the same for the foot or the club.

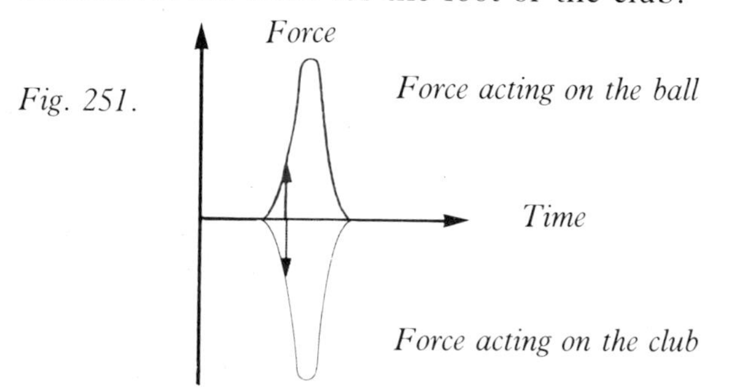

Fig. 251.

(c) Friction

Friction is the resistance to motion of two objects or surfaces that touch. Friction is which depends on the fact that small irregularities between shoe and ground, racket and ball, skis and snow, etc., must be evened out if the objects are to glide against each other.

The magnitude of the force of friction depends on the kind of materials in contact with each other (the roughness of their surfaces, μ) and on how hard the two surfaces are pressed together *(N)*.

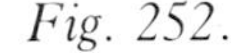

Fig. 252.

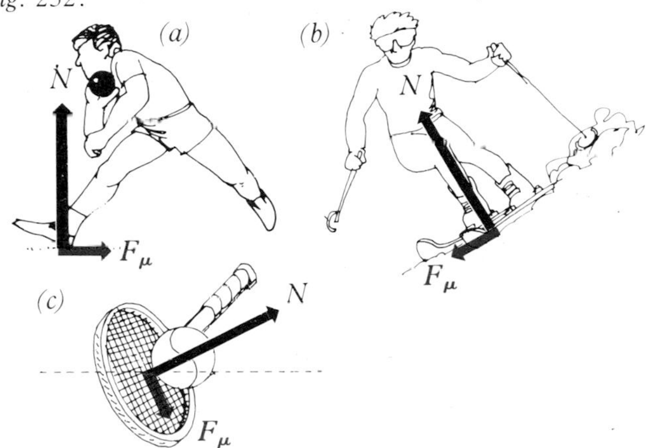

Example 1 (Figure 253). If an object is exerting a force of 100 N on the ground, and a force of 20 N is needed to make it glide along the surface, it is said that the coefficient of friction (μ) is 20/100 = 0.2.

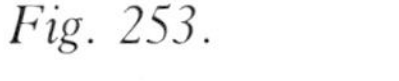

Fig. 253.

Example 2 (Figure 254). If a skier who weighs 60 kg glides along a horizontal surface — the entire time being resisted by a force of 30 N — it is said that the coefficient of friction is 30/600 = 0.05.

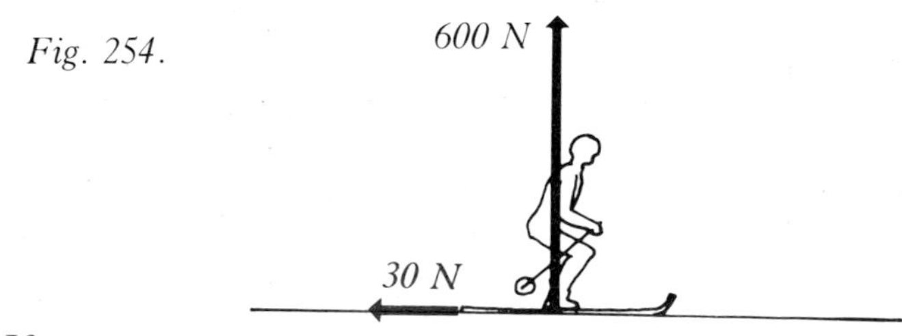

Fig. 254.

If $\mu = 0$, then there is no friction.
If $\mu = 1$, then we may agree that the object is fixed.

(d) Air resistance (F_a)

Fig. 255.

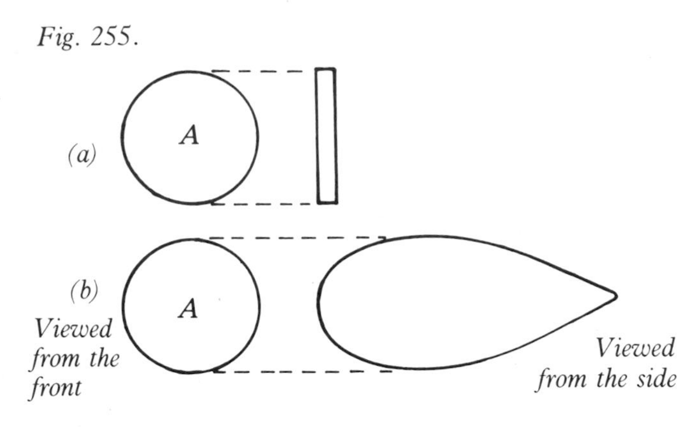

The resistance met by a body moving through air or water depends on the size of its surface area *(A)* directed forward, its velocity *(v)* and its form *(c)*.

Two bodies *a* and *b* may have the same area when viewed from the front, but very different streamline forms when viewed from the side.

Fig. 256.

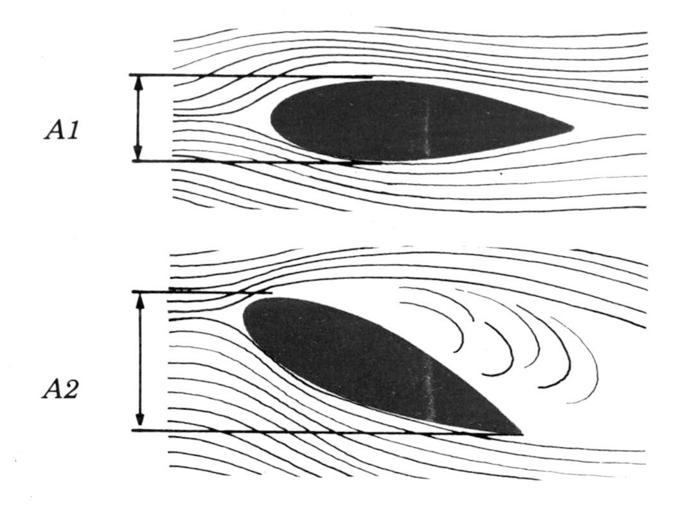

A change of body position often leads to a change in both *A* and *c*. The equation of air (water) resistance has been established by experiment: $F_a = A \times v^2 \times c$. Thus, if the area is doubled, the force of resistance will be doubled; doubling the velocity gives four times the force of resistance and improving the streamline form reduces the constant *(c)*, i.e. less air resistance.

The aim in such sports as long jumping, ski jumping, skating and, in particular, downhill skiing, is to reduce air resistance. In swimming, it is especially important to reduce water resistance.

Fig. 257.

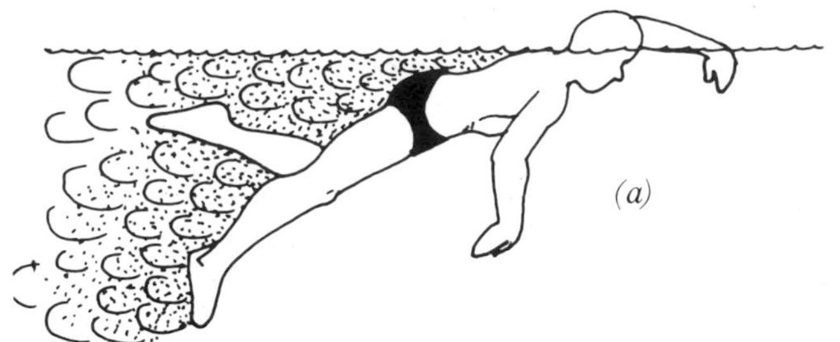

Both the area (A) and the streamline form (c) are greater in (a) than in (b).

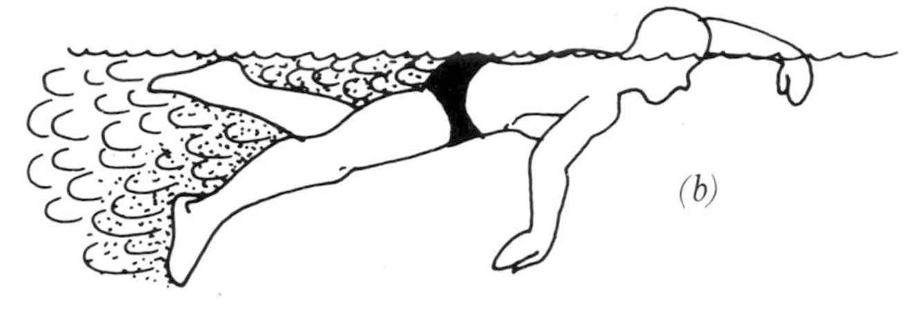

Fig. 258.

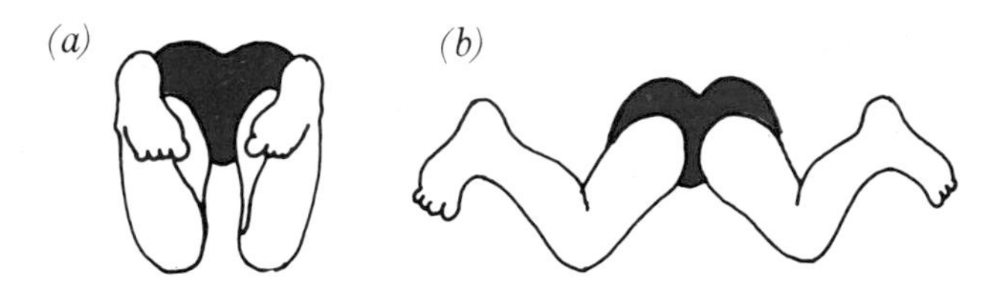

The breast stroke kick assumes little area (A) while the heels are drawn in towards the buttocks.
Large area when the legs kick backwards.

Internal Forces
(e) Muscle forces

Muscle forces are counted among the so-called internal forces. In order to resist the forces of gravity and friction, or to increase the normal force, the body uses muscle forces. Muscle forces affect the origin and insertion of a muscle to exactly the same extent (Fig. 259), i.e. with equal but opposing forces. The size and structure (position of the centre of gravity) of the body parts A and B are the factors which determine what happens. Also important is whether A and B are free to move, or whether one of them is in some way "fixed".

Fig. 259.

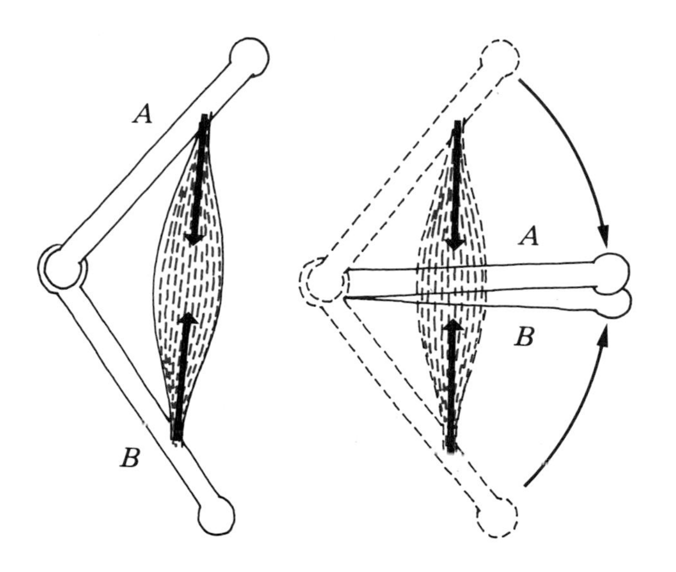

Example 1 (Figure 259). If A and B are identical and both are free to move, they will come together much like a folding jack-knife. This is roughly what happens when a person jumps up with a straight back and flexes his hip joint. The iliopsoas muscle pulls his legs up and his trunk down. His legs are a little lighter than his trunk so they travel a longer distance. It is impossible to move the legs alone.

Fig. 260.

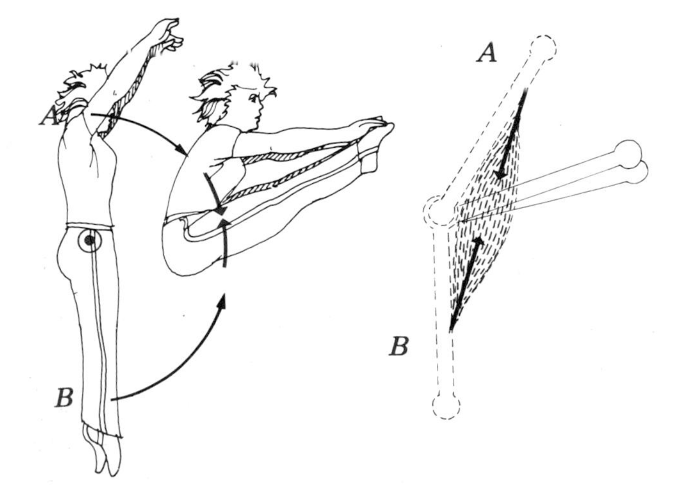

Example 2 (Figure 260). If B is held stationary, only A will move.

Fig. 261.

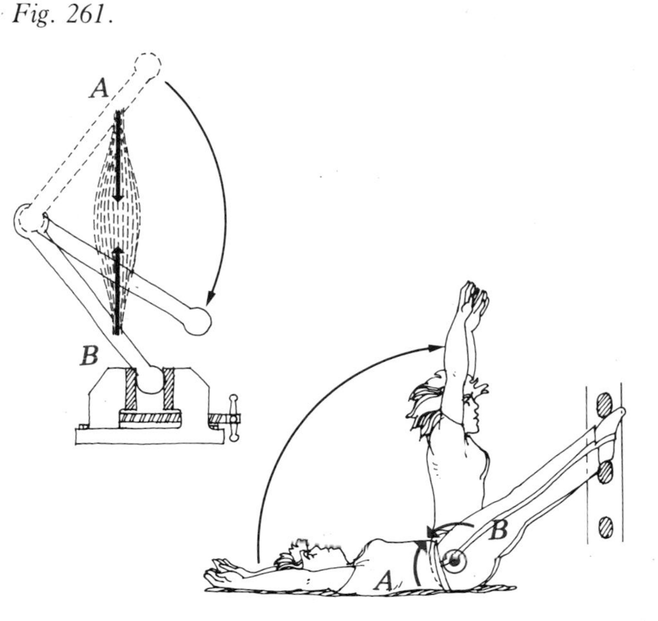

Example 3 (Figure 261). If the weight of the parts differ, the lighter of the two moves the farther. The diagrams below present different situations where both parts rotate in opposition to each other at a joint.

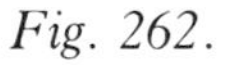

Fig. 262.

(a)

Figure 262. If the trunk is rotated in one direction, the skis try to turn in the opposite direction.

(b)

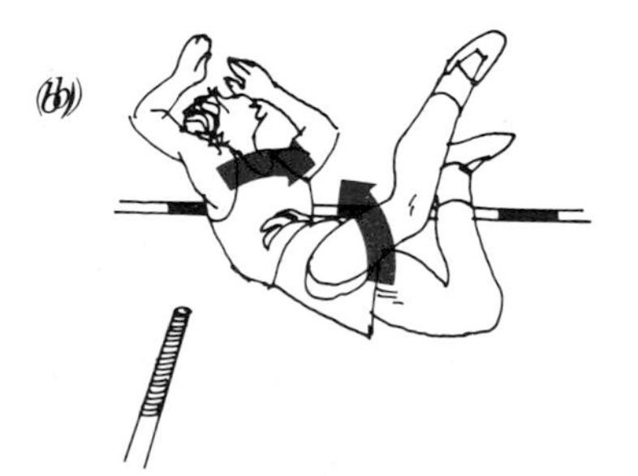

In pole-vaulting, the upper part of the body is lifted over the bar by extending the hip joint and kicking backwards with the legs.

Fig. 262.

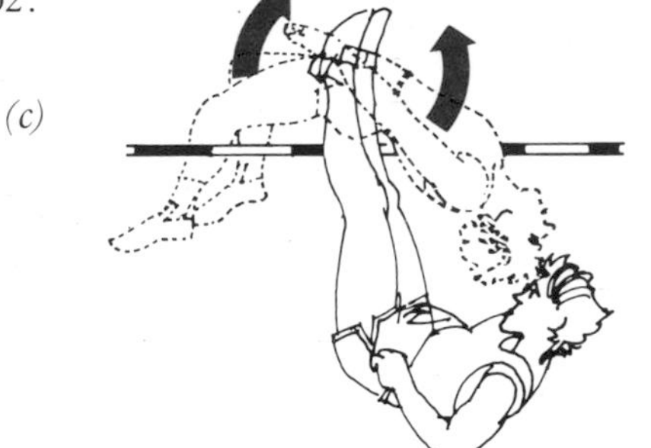

To avoid knocking the bar off with the legs in the flop style high-jump, the hip joint is powerfully flexed once the legs have passed over the bar. By rotating his arms and legs backwards as far as possible, the high-jumper prevents the upper part of his body from rotating, and thus landing on his neck.

By thrusting his trunk forward and continuing the rotation of his arms, the long-jumper causes an opposing swinging movement of his legs in the landing phase of the jump.

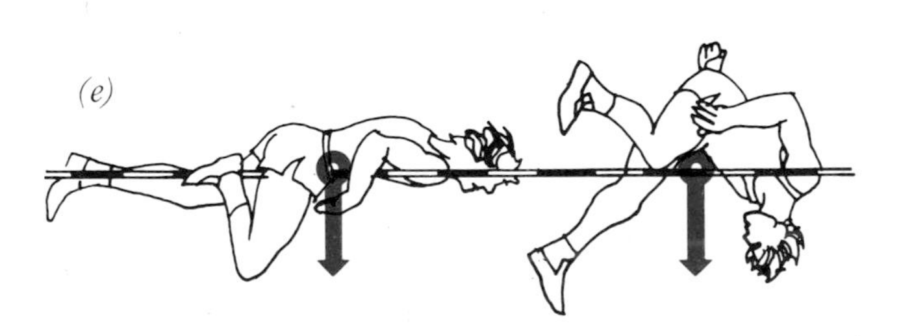

One way to avoid knocking the bar off with the trailing leg in the straddle style high-jump, is to pull the knee up by strongly flexing the hip joint once the body has cleared the bar.

When preparing to throw (the discus, or whatever), the skilful athlete pulls his "non-throwing arm" well back as shown in Figure 262 (f). Before actually hurling the object this movement is blocked. The arm is thrust in the opposite direction in order to increase the rotation of the trunk which in turn increases the velocity of the throwing arm.

(f) Tendon and ligament forces

These are passive inner forces. They are brought about by muscle or external forces, i.e. the tendons and ligaments cannot themselves produce force. When subjected to severe stress from without, a ligament may rupture (Figure 263).

Fig. 263.

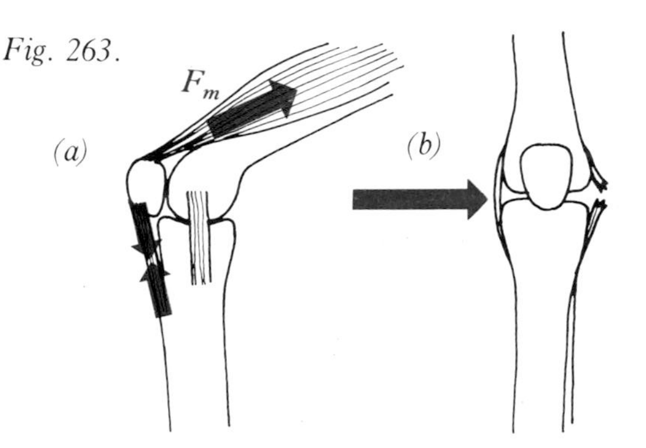

II. Centre of Gravity

The centre of gravity of a body is the point at which it could, in theory, be suspended and remain in equilibrium regardless of its position.

It is easy to discover the centre of gravity in a symmetrical body; it is more difficult to do so in an asymmetrical one. Nevertheless, a trainer should be able to discern just where the centre of gravity of an athlete's body lies in any given situation. By doing so, he can correctly instruct an athlete on how to execute exercises.

Fig. 264.

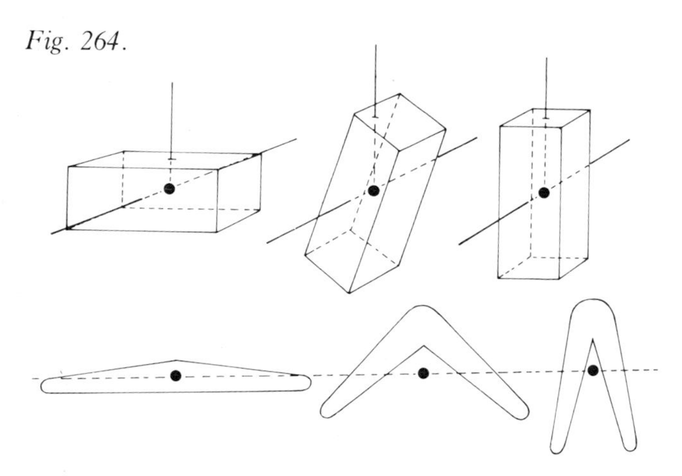

Fig. 267.

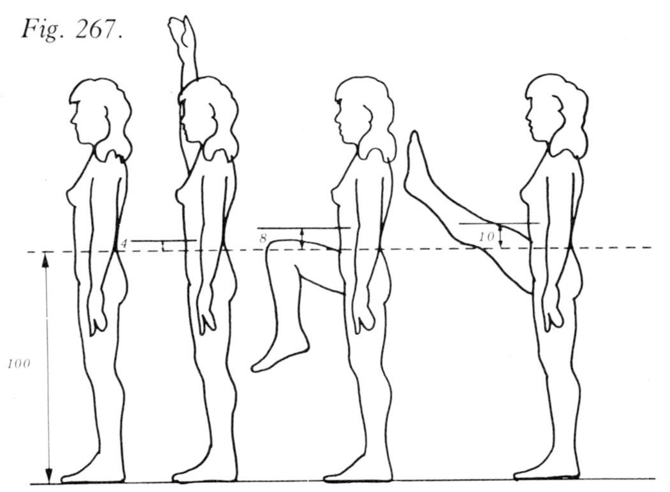

Figure 267 demonstrates how the centre of gravity varies in relation to the ground.

If a person jumps up so that her centre of gravity reaches 150 cm above the ground (Figure 268), a change in body position will not affect the height of the centre of gravity in relation to the ground. Once the body has left the ground its position on the way up will not affect the maximal height of the centre of gravity.

Fig. 265.

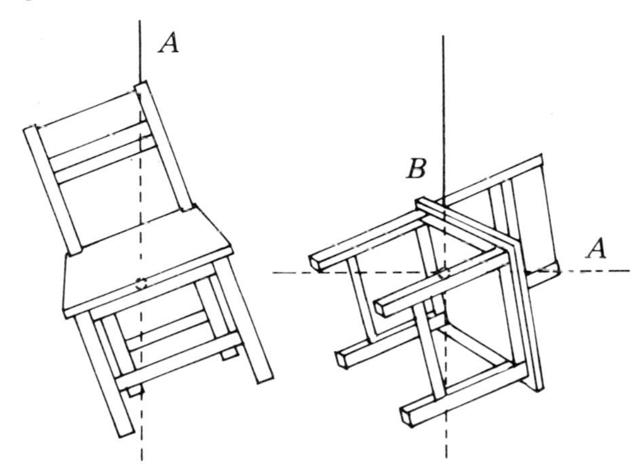

The centre of gravity of a rigid body can be determined by experiment in the following way: hang the body freely at an arbitrary point A. The centre of gravity will lie on the vertical straight line passing through A. Hanging the body from some other point B gives another straight line on which the centre of gravity lies. Clearly, the centre of gravity is the intersection of the two lines (Figure 265).

Fig. 266.

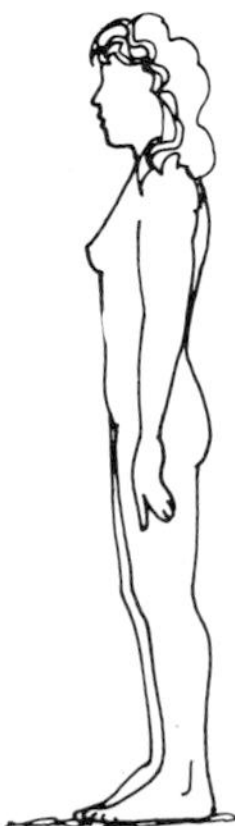

The centre of gravity of a human body standing in the anatomical position (Figure 266), lies approximately in line with the navel a few centimetres in front of the third lumbar vertebra. Naturally, the location of the centre of gravity changes with changes in body position (e.g. when an arm or leg is raised). If the centre of gravity in Figure 266 lies 100 cm above the ground it will rise about 4 cm when an arm is raised, 8 cm when both arms are raised, 8 cm if the body stands on its toes, etc. (see Figure 267).

Fig. 268.

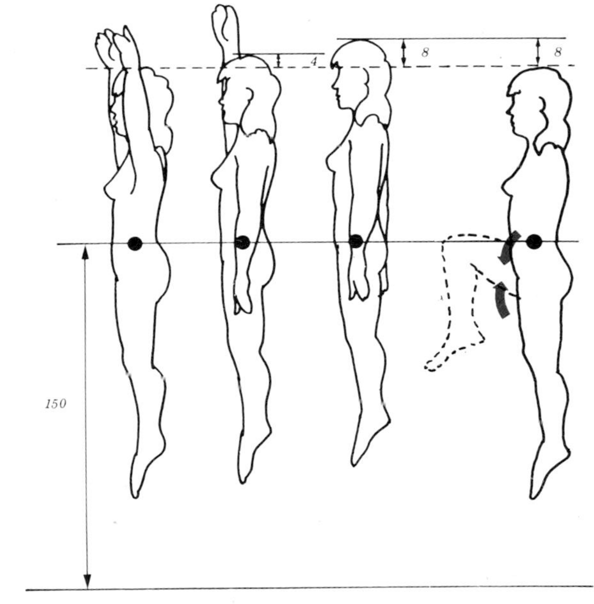

Figure 268 shows that, although the centre of gravity reaches the same height in all four jumps, the height of the hand and head above the ground varies with the final position. If both arms are lowered, the body is raised 8 cm. If one leg is raised, the rest of the body will be lowered 8 cm (cf. diagrams overleaf). The explanation is quite simply that the muscle group which lifts the leg (e.g.

iliopsoas) affects the rest of the body with an equal but opposing force, i.e. pulls the body down.

Fig. 269.

The pattern of movement in such activities as the jump shot in handball, shooting in basketball, smash in volleyball, kick in football, clearing the bar in high jump and pole vault, hurdling, etc., is built on the principles outlined above.

Notice that by keeping the left arm and both legs straight at the moment of shooting, the right hand will be raised to a maximal height (Figure 269).

Fig. 270.

By holding both arms and right leg as high as possible, the left leg will reach the ground at the earliest possible moment, and the hurdler can begin to run again (Figure 270).

Fig. 271.

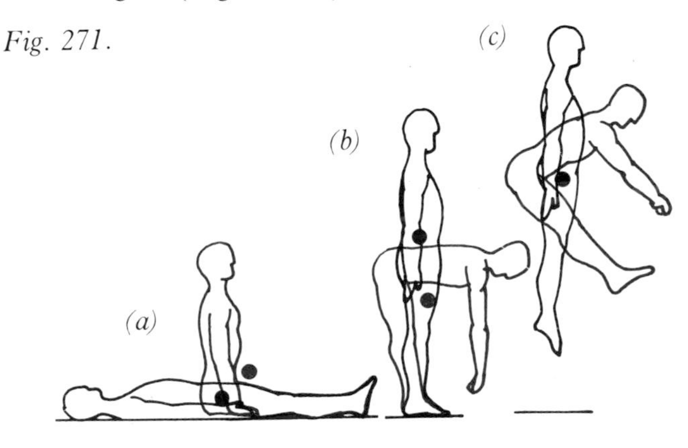

Figure 271 further illustrates variations in centre of gravity location. Hip flexion causes the centre of gravity to vary in three different ways depending on the situation.

Figure 272(a) shows where the centre of gravity lies in a certain part of the body expressed as a percentage of its length. For example, the centre of gravity of the arm lies at a point which — measured from the shoulder downward — corresponds to 40% of its length. The figures shown in (b) give the mass of each body part expressed as a percentage of total body weight. For example, the weight of the head is 7% that of the whole body.

Fig. 272.

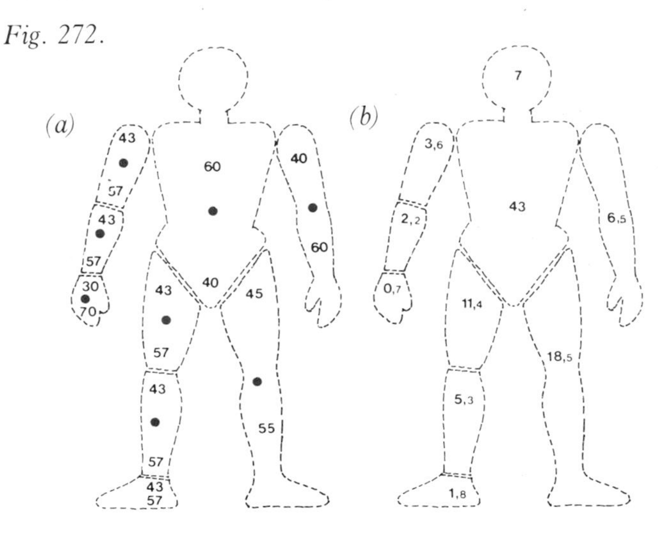

The diagrams and calculations in Figure 273 show how we can make a total assessment of different exercises, once we know the location of the centre of gravity of a body, its moment of force and the function and position of its muscles.

Fig. 273.

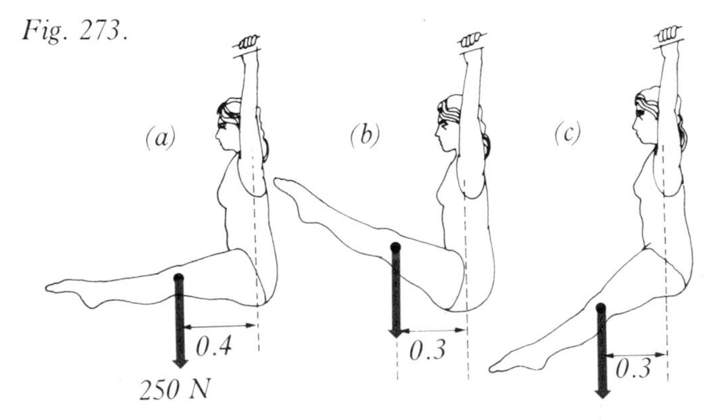

Suppose a person hangs so that the centre of gravity of her legs (45% of the leg's length from the hip) lies 0.40 m directly out from her hip in (a), 0.30 m in (b) and 0.30 m in (c). If her legs weigh 25 kg (18.5% of her total body weight per leg), then the gravitational force acting on them will be 250 N. The moment of force of (a), (b) and (c), will then be:

(a) $M = F \times d$
$M = 250 \times 0.40$
$M = 100$ Nm.

(b) $M = F \times d$
$M = 250 \times 0.30$
$M = 75$ Nm.

(c) $M = F \times d$
$M = 250 \times 0.30$
$M = 75$ Nm.

Fig. 274

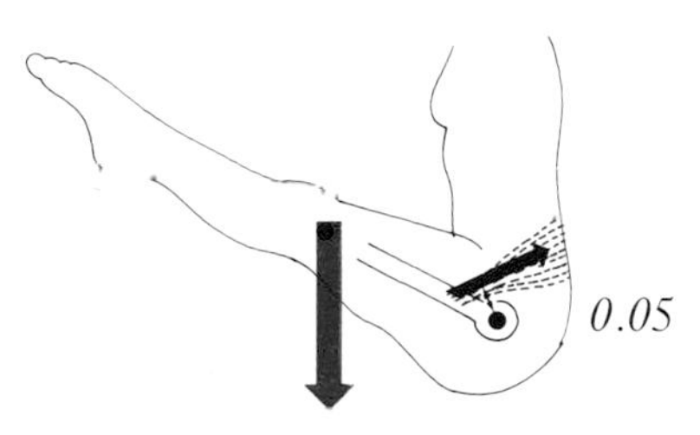

Suppose the iliopsoas is the muscle holding the legs, and that its lever arm is 0.05 m (Figure 274). Then,

(a) $F \times 0.05 = 100$
$F = 2000$ N.

(b) $F \times 0.05 = 75$
$F = 1500$ N.

(c) $F \times 0.05 = 75$
$F = 1500$ N.

Fig. 275.

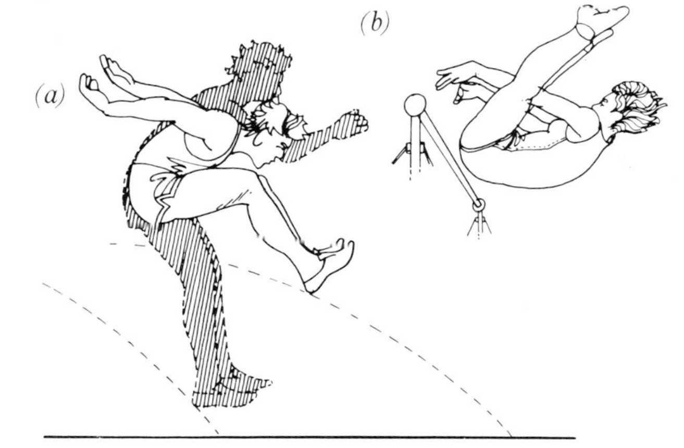

All those who have tried position (b) (Figure 273) know that it is almost impossible, despite the fact that the muscle force in (b) and (c) is the same according to the calculations. However, the iliopsoas is longer in (c) than in (b), and thus more capable of exerting sufficient force. In (b) the distance between the origin and insertion of the muscle is perhaps less than 50% of its resting length. This means that the muscle cannot develop any force at all (p.18). Another reason why it is "impossible" to hold the legs in position (b) is that the hamstrings (p.54) are so drawn out that they try to pull the legs down with great force much like an elastic band strives to return to its resting length.

In the situations shown in Figure 275, the interplay between hip flexion strength and hamstring flexibility, is of decisive importance for the pattern of movement.

If the legs or upper body are folded forward too early, the legs cannot be held in the best position for landing. Instead, the body "opens" so that the feet land too soon.

The following observation is made to illustrate further the centre of gravity concept and its importance. When a person lifts an object, different muscles will be engaged depending on how he moves the object.

Fig. 276.

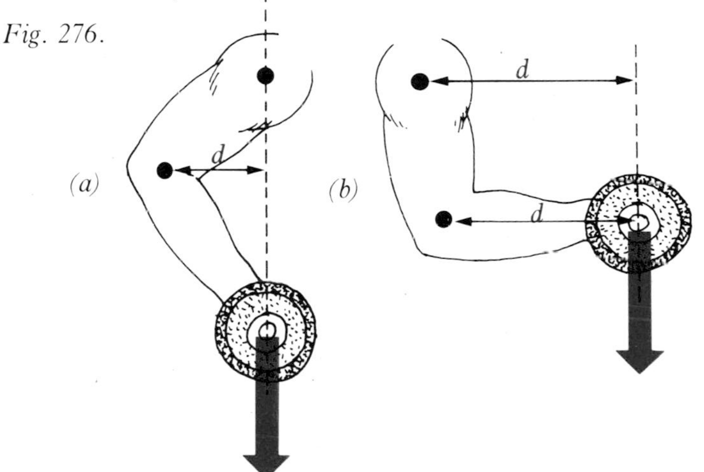

If he lifts as in Figure 276 (a), the work load to which his elbow flexors are subjected, will be less than if he lifts as shown in (b) (shorter lever arm *d*). The work load to which the shoulder joint is subjected is almost zero in (a) but very large in (b).

Fig. 277.

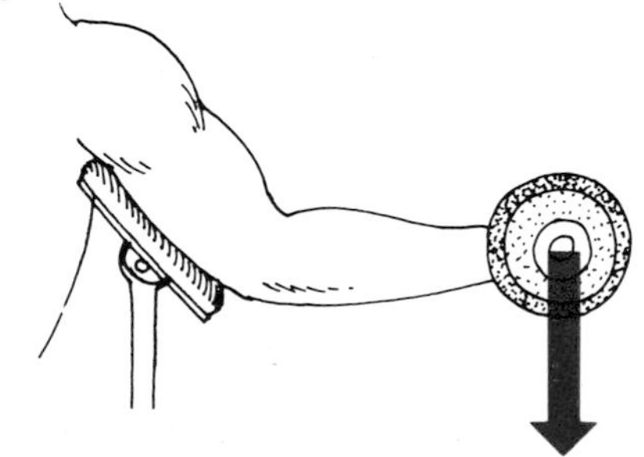

Training the elbow flexors for strength using a support for the upper arm (Figure 277), forces the elbow flexors to work the "heavy way" and, at the same time disengages the shoulder muscles.

Fig. 278.

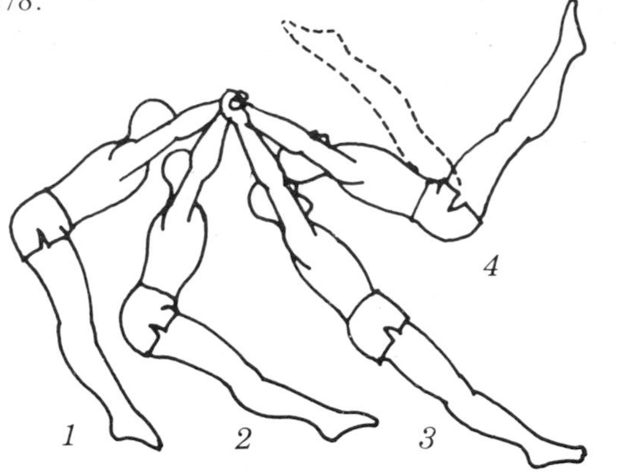

In the forward swing on the horizontal bar (Figure 278), it is important to be so arched at the hip that the legs can be lifted quickly as the gymnast approaches the forward limit of his swing. Thus, his hips should be kept as straight as possible in positions 1-4. If his hips are piked at 1–2, then his legs will start falling (i.e. they will be on their way down when it is time for them to be lifted).

Fig. 279.

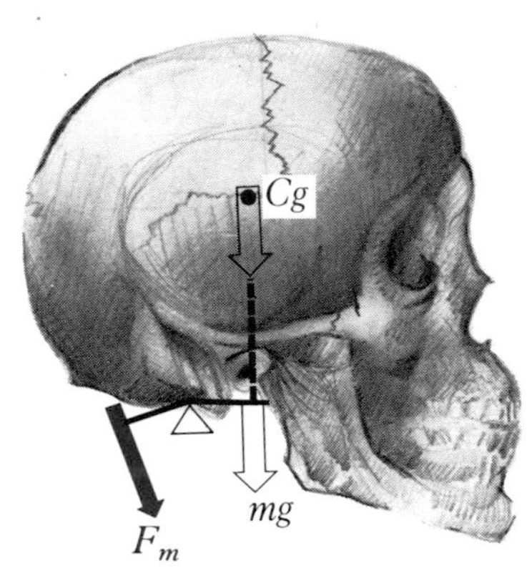

Fig. 280.

Fig. 281.

Fig. 282.

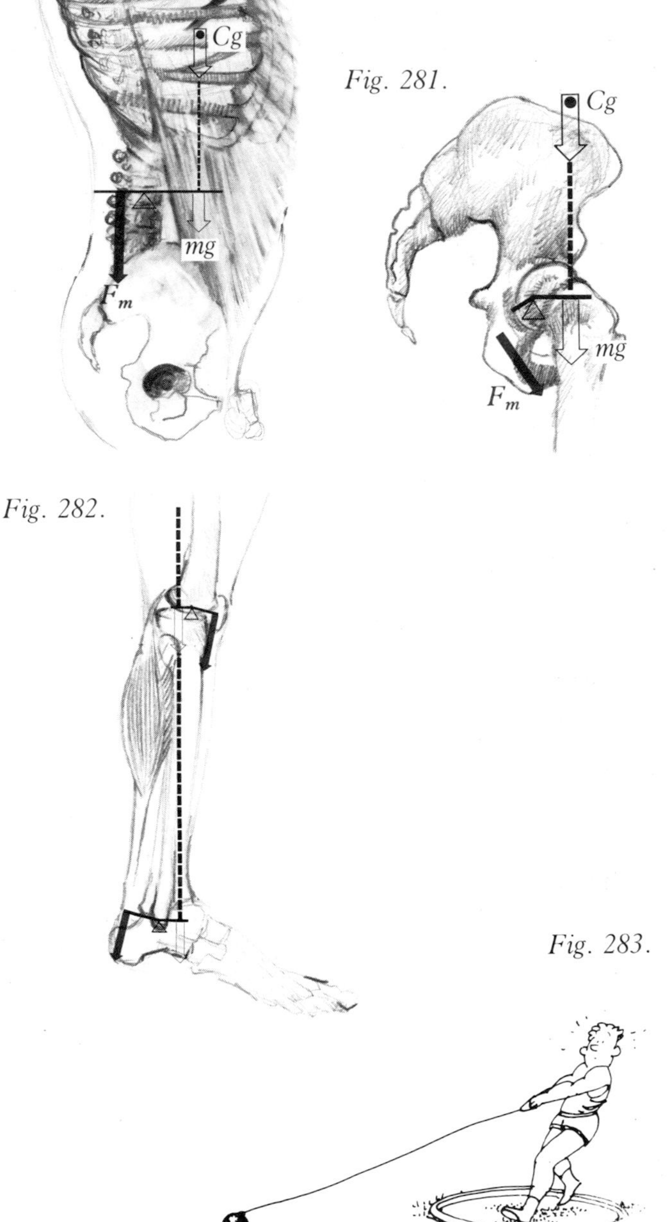

Fig. 283.

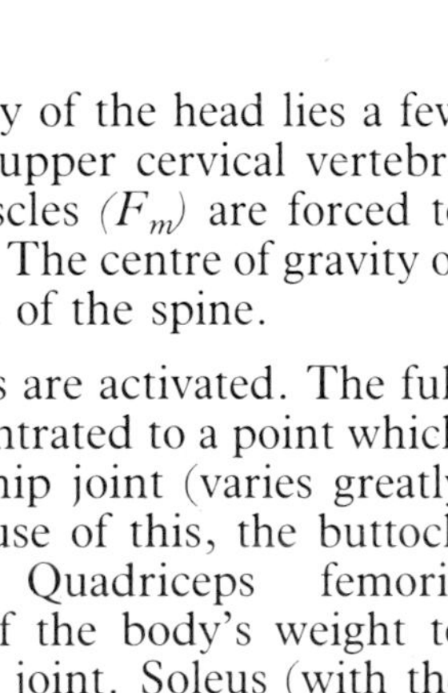

Because the centre of gravity of the head lies a few centimetres in front of the upper cervical vertebra (Figure 279), the neck muscles (F_m) are forced to work statically to hold it up. The centre of gravity of the upper body lies in front of the spine.

Therefore, the back muscles are activated. The full weight of the trunk is concentrated to a point which lies a little in front of the hip joint (varies greatly between individuals). Because of this, the buttock muscles are engaged. Quadriceps femoris counteracts the tendency of the body's weight to produce flexion at the knee joint. Soleus (with the help of gastrocnemius counteracts the body's tendency to fall forward at the ankle joint. The function of all these muscles is to maintain the body's upright position. They are called postural muscles (Figures 280 and 281).

When a particular part of the body is to be moved, it is very important to be aware of the location of the centre of gravity. When an object is to be swung or whirled around, it can be shown that it is easier to initiate the movement when the radius of rotation is small. A hammer whose wire has a radius of 0.5 m is easier to swing than one whose wire is, say, 1 m. A hammer with a 3 m long wire sling would be completely unmanageable.

If a person wants to swing his leg quickly forward while running, it is important that he keep the swing radius as small as possible (Figure 284). This can be accomplished by pulling the heel up towards the buttock (3) as soon as the foot leaves the ground (1). The leg is then swung forward with its centre of gravity as close as possible to the hip.

Fig. 284.

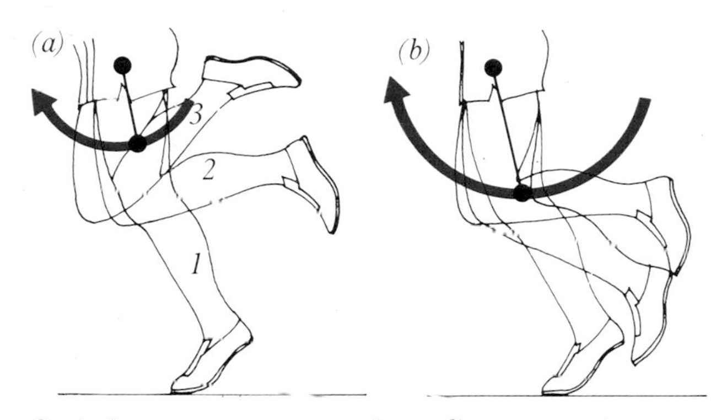

Sprinting *Long distance running*

This type of running is typical for sprinters. Long distance runners, on the other hand, do not waste energy by lifting the heel so high (which requires strength and speed of the hamstrings). Instead, they have time to allow the leg to swing forward aided by the force of gravity and the hip flexors.

The arm is bent in the initial stage of almost any throwing or hitting event (Figure 285). As long as the shoulder muscles are accelerating the arm, the centre of gravity of the arm should be close to the shoulder. When throwing an object, the muscle groups are activated in the order: abdomen, shoulder, elbow, wrist.

Fig. 285.

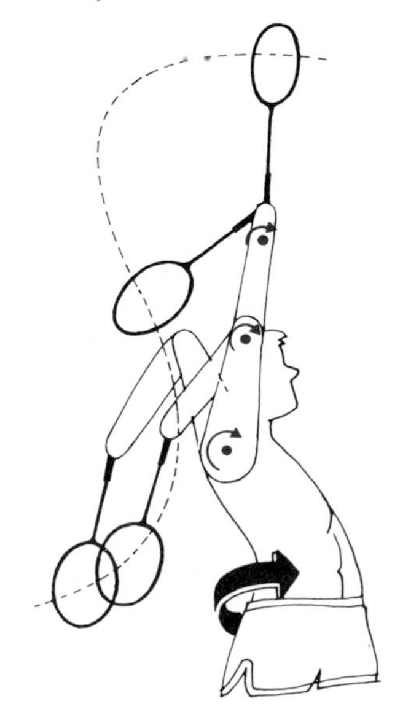

The pattern is the same when kicking a football (Figure 286). The muscles are then activated in the order: abdomen, hip, knee.

Fig. 286.

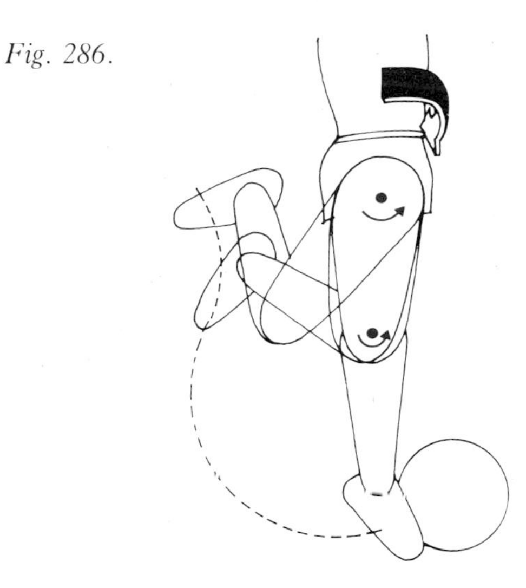

On p.99 it was pointed out that when two parts of a body of equal weight are brought together, they move symmetrically. This is only true when the distance between the centre of gravity and the joint is the same for both.

Fig. 287.

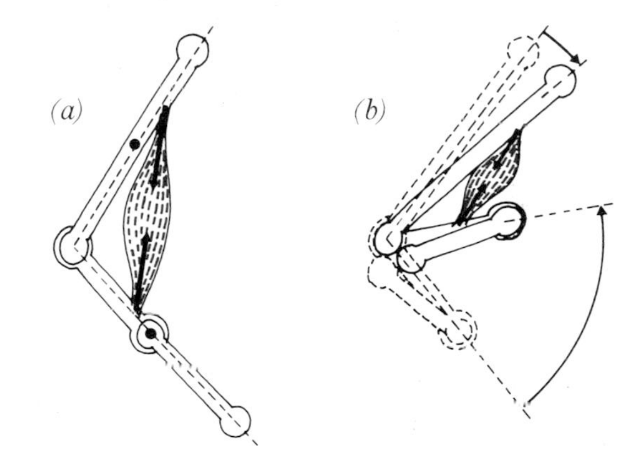

If this is not the case, the part whose centre of gravity is farthest away from the joint will move through a smaller angle. This phenomenon is often used in athletics.

Thus, for effective movement at a joint, the parts of the joint should be arranged in such a way that the centre of gravity of the part to be moved lies close to the joint, while the centre of gravity of the other lies far from it.

The following four examples [Figure 288 (a)–(d)] illustrate this principle when the legs are swiftly swung forward.

Fig. 288.

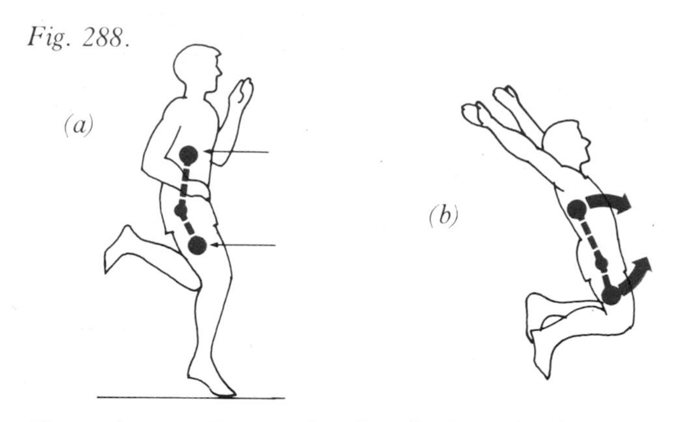

Example (a). By maximally flexing the knee, the centre of gravity of the leg is brought to a point near the hip joint. This makes the leg easy to swing forward. The centre of gravity of the upper body should, in principle, lie far from the hip joint, which implies that the arms should be relatively well flexed.

Example (b). Before swinging his legs forward a long jumper first bends them maximally behind him (Cg near the hips) and, at the same time, stretches his arms high above his head (Cg far above the hips).

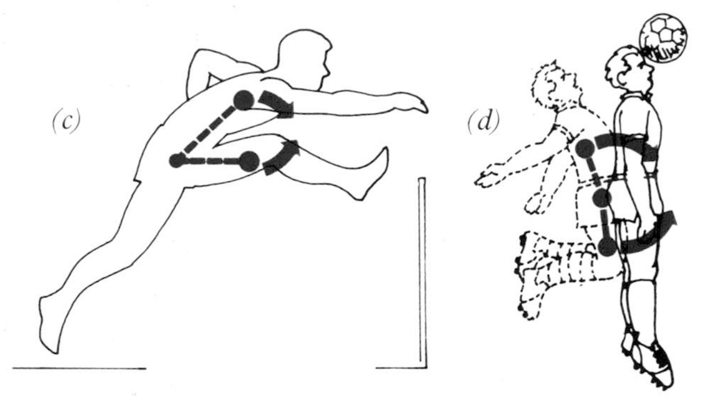

Example (c). If a hurdler wants to make the most of the muscle force of his hip flexors, he should hold both his arms as far from his hips as possible (right arm straight, left arm bent). His right leg should remain bent on the way up and he should wait as long as possible before straightening it.

Example (d). When a footballer heads the ball, his whole upper body should be moving quickly. Here, the centre of gravity of his upper body should lie near his hips (arms by his sides). The moment the ball is hit, the centre of gravity of his legs should lie far from his hips (straight legs). Imagine how ineffective a header would be if made from a position corresponding to that shown in example (b), above.

The following five examples [Figures 289 (a)–(c) and 290 (a) and (b)] show how to arrange the body when making arm movements.

Fig. 289.

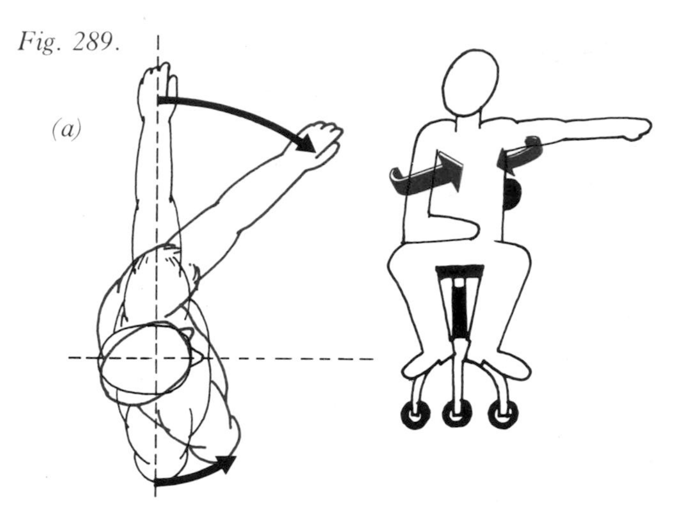

Example (a). If a person moves his arm in an arc in front of his body, his body will respond by rotating in the opposite direction. If his arm is bent, his body moves less. This and other similar effects can easily be observed with the aid of a mobile platform (turntable) or swivel chair.

Example (b). To roll his trailing leg (left) over the bar in the straddle style high jump, the high-jumper should pull his right arm in the opposite direction. Notice how the right arm changes position and rotates up behind the back in diagrams 2 and 3, below. The arm should be held straight and a long way from the body.

Example (c). A straight arm that counters the rotation of the body is more effective for throwing than a bent arm that merely pulls in towards the body.

Fig. 290.

(a)

Example (a). When landing, the action of pulling the arms backwards and downwards causes the body to rotate in the opposite direction (or, the arm movements reduce the body rotation).

As a result, the feet land further to the left in (a) than would be the case without arm movements.

Example (b). If the arms are pulled down in front of the body, the feet will land further to the right in (b). This is due to the opposing arm movement. The muscles that pull the arms down arise from the chest and therefore must draw the body up. If the arms are kept straight, the opposing movement of the body is greater than if the arms are bent. If the arms are pulled outwards and downwards the body's rotation is not affected.

III. Moment of Inertia

The previous examples show that the further out from the centre of rotation the centre of gravity of a particular part of the body lies, the greater is the force required of the muscles to set it in motion. Of course, the mass (weight) of the body part is also important. The more the body part weighs, the more force is needed to set it in motion (or stop it).

In physics, the term moment of inertia is used to describe not only the weight of a body but also how

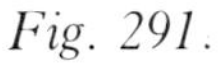

Fig. 291.

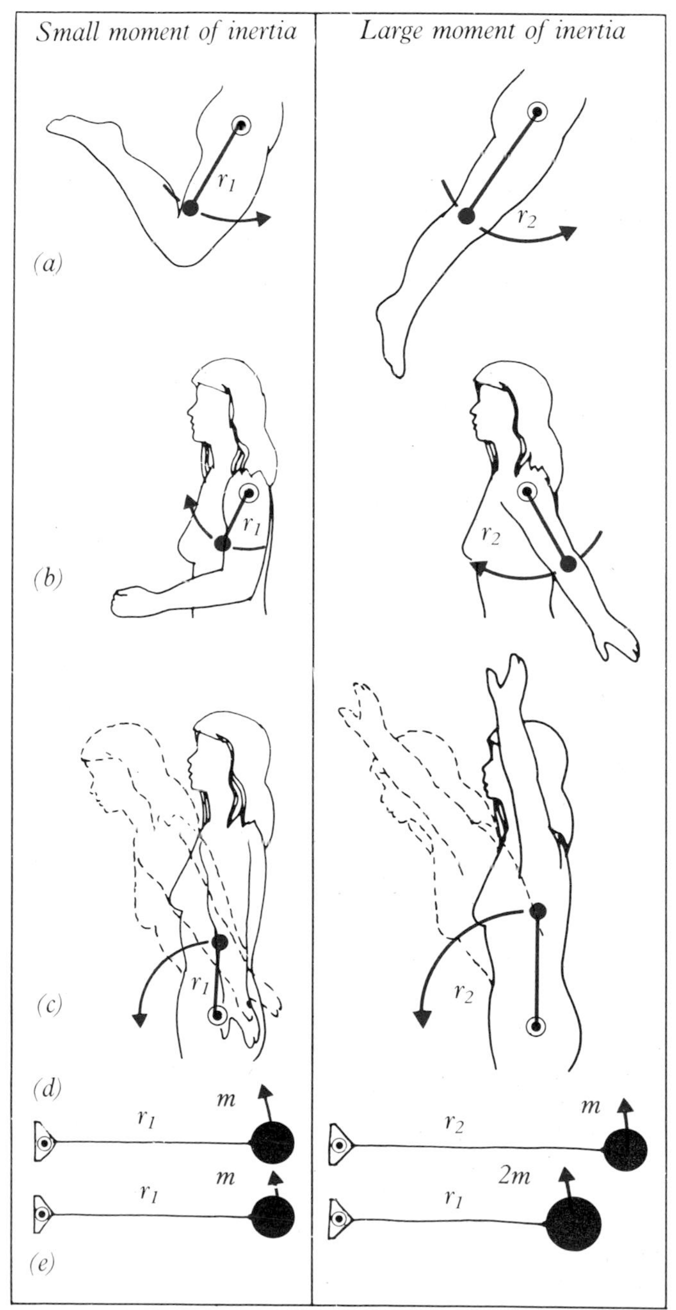

far from the axis of rotation the centre of gravity of the body lies. The figures in the left-hand column in Figure 291 have less moment of inertia than their counterparts in the right-hand column.

In physics the letter I is used to symbolise moment of inertia. In order to determine the moment of inertia of a body we must know its mass (m) and its rotational radius (r). The moment of inertia is

$$I = mr^2.$$

Fig. 292.

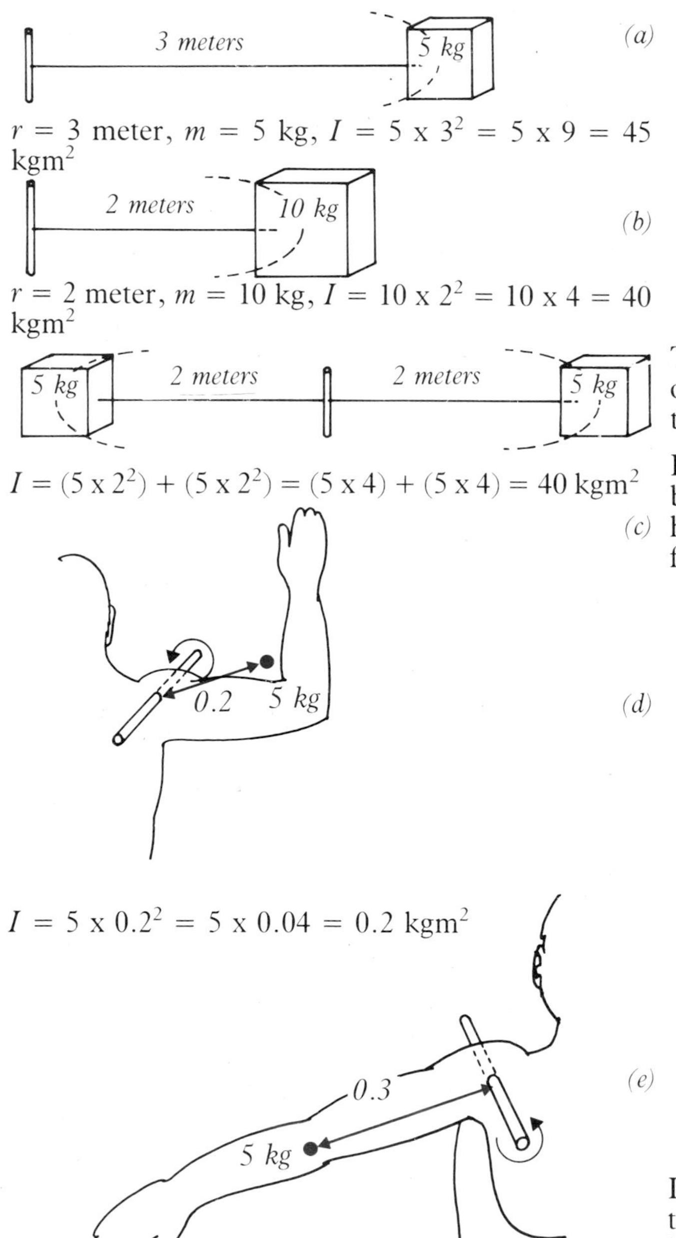

$r = 3$ meter, $m = 5$ kg, $I = 5 \times 3^2 = 5 \times 9 = 45$ kgm^2

$r = 2$ meter, $m = 10$ kg, $I = 10 \times 2^2 = 10 \times 4 = 40$ kgm^2

$I = (5 \times 2^2) + (5 \times 2^2) = (5 \times 4) + (5 \times 4) = 40$ kgm^2

$I = 5 \times 0.2^2 = 5 \times 0.04 = 0.2$ kgm^2

$I = 5 \times 0.3^2 = 5 \times 0.09 = 0.45$ kgm^2

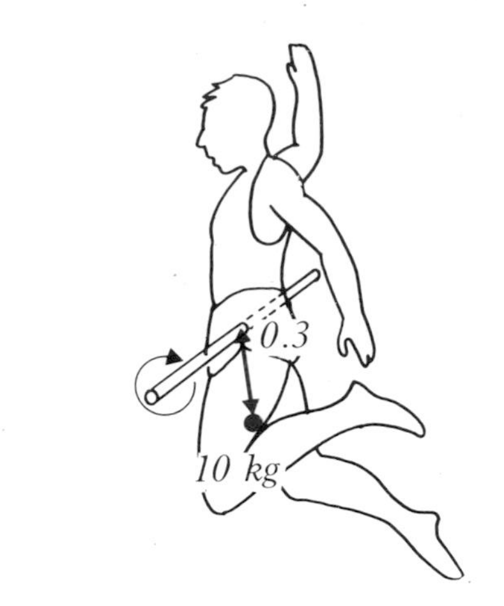

$I = 10 \times 0.3^2 = 10 \times 0.09 = 0.9$ kgm^2

The underlined numbers in Figure 292 give an idea of how great a force must be exerted in order to set the different bodies in motion.

If we want to calculate the moment of inertia of a body rotating about its own centre of gravity (which happens in all flights through the air), we do the following.

Fig. 293.

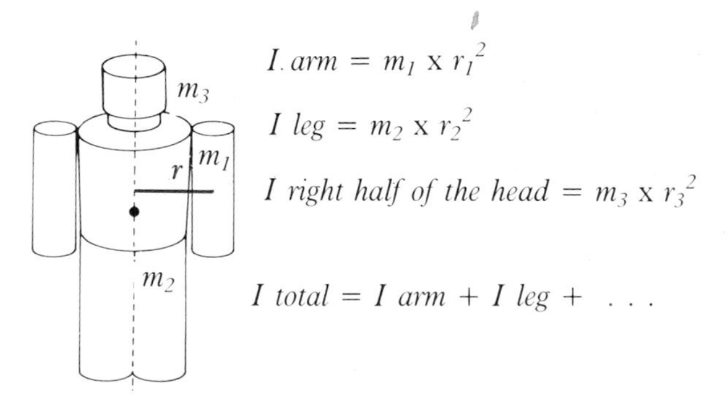

It has been shown with this type of calculation that the moment of inertia for rotation about a body's longitudinal axis varies as the body takes on different positions. In Figure 294 variations in body position are compared with the anatomical position (I).

Fig. 294.

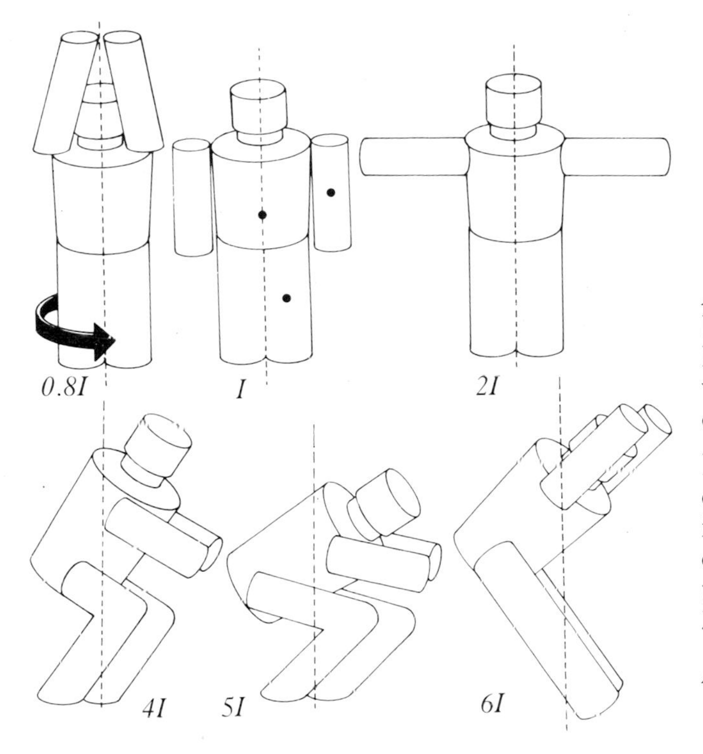

The moment of inertia will increase as more body segments move away from the longitudinal axis of rotation, i.e. as the distance *(r)* to the axis increases.

The moment of inertia for forward and backward rotation (vault) varies roughly according to the following data.

Fig. 295.

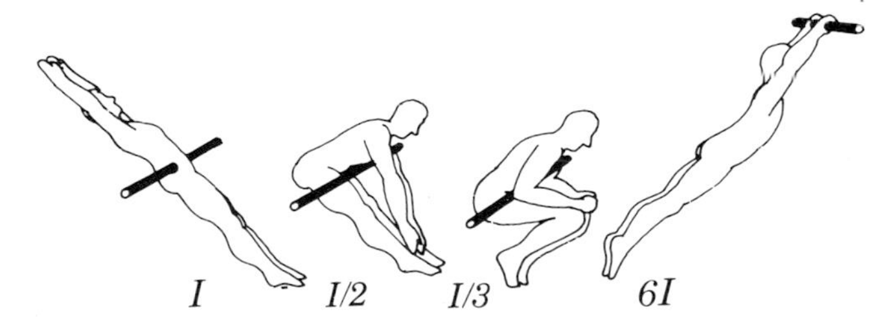

It has been shown by experiment that when a body is made to rotate by some force and thereafter left to rotate without being influenced by other external forces, its rotational velocity (*w*) will depend on its position. It is said that moment of inertia x rotational velocity does not change. *I* x *w* = constant. This has the following consequences:

Fig. 296.

Example 1. If a person who is rotating with his arms by his sides *(I)* at a certain rotational velocity (*w*) lifts his arms out, he will reduce his rotational velocity by half (*w*/2). The moment of inertia will be changed from *I* to *2I* by this change in his position.

If he sinks down to a half sitting position, *I* will be changed to 4*I*, and hence his rotational velocity (*w*) is reduced to a quarter (*w*/4). An example of such changes being brought about is the execution of a pirouette: jumping – pirouette – landing (Figure 296).

Fig. 297.

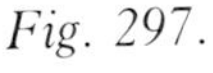

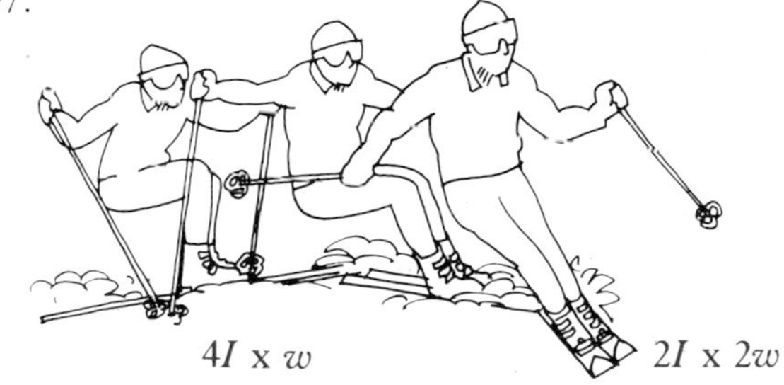

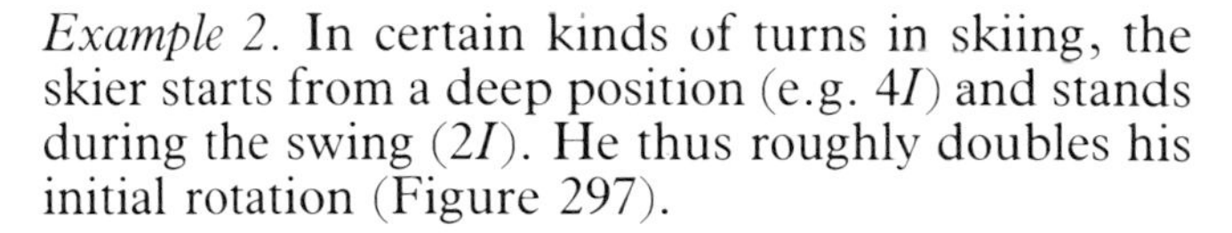

Example 2. In certain kinds of turns in skiing, the skier starts from a deep position (e.g. 4*I*) and stands during the swing (2*I*). He thus roughly doubles his initial rotation (Figure 297).

Fig. 298.

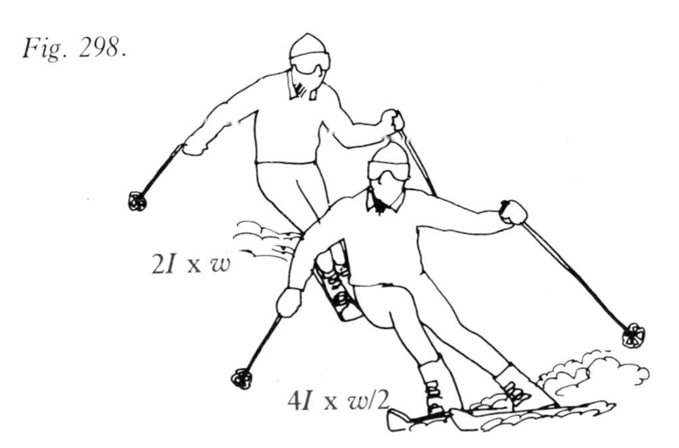

Example 3. When coming out of a turn, the skier can reduce his rotational velocity to about a half by dropping down to a deeper position (from 2*I* to 4*I*). Hence, he will find it easier to steer straight again (Figure 298).

Fig. 299.

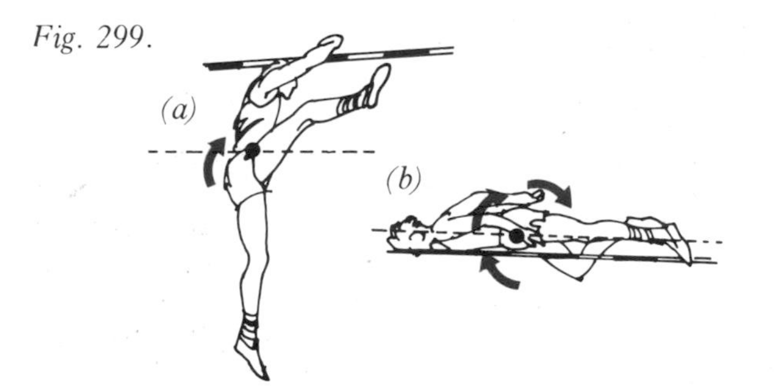

Example 4. In the take off of the straddle style high jump, the body reaches a position of great rotational inertia about an axis which is parallel to the bar [Figure 299 (a)]. If the jumper cannot quickly change his position, he risks knocking the bar off with his trailing leg. This is because his rotation is very slow in this position. When he comes to straddle the bar with his trailing leg, it is important that his rotational velocity is at a maximum, i.e. the moment of inertia should be minimal. He does this by keeping his body straight and his arms close to his sides [Figure 299 (b)].

Fig. 300.

Example 5. The flop style high jumper rotates backwards with greater speed if he arches his back than if his body is straight. He has better stability against sideward rotation if he holds his arms out than if he keeps them close to his body (Figure 300).

Fig. 301.

Example 6. A thrower increases the total rotational velocity of his body (and thus the velocity of his throwing arm) in the following way. In the final phase of the throw, he rises from a relatively low position to one which centres all the parts of his body — except his throwing arm — around the rotational axis (Figure 301).

Fig. 302.

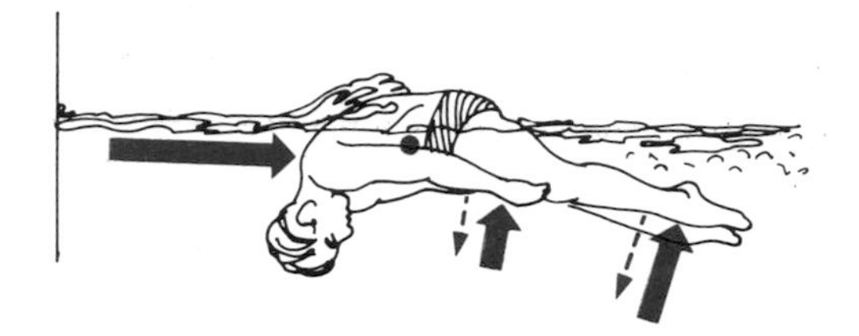

Example 7. A swimmer starts a front crawl flip turn by flipping his legs down against the water (which resists by pressing against him as shown by the arrows in Figure 302). His hands also press in the "wrong direction". By piking at the hip, he makes the water press against his back. These three measures have given him a certain rotational velocity (w). At this point, he folds up maximally, thereby doubling the rotational velocity he acquired.

Fig. 303.

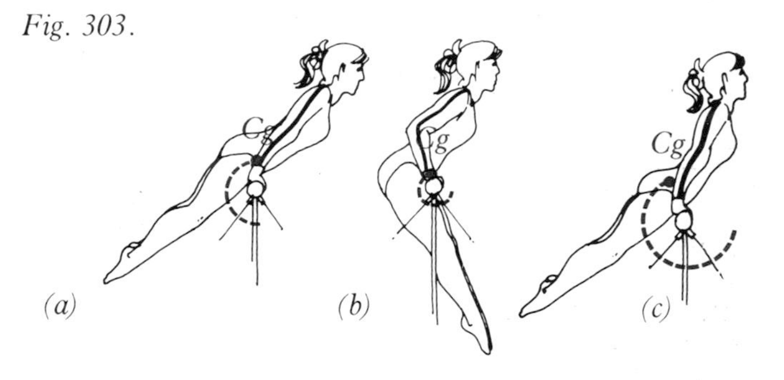

Example 8. A gymnast who wants to rotate about a bar (wheel swing) has greater difficulty in swinging around if her body is kept completely straight [Figure 303 (a)] than if there is a slight piking at her hips (b). By piking slightly she shifts her centre of gravity to a place outside her body, i.e. near the bar. If she tries to swing around by pulling her head backwards, that is by arching slightly, her difficulties will only increase (c).

APPLICATIONS

This chapter concludes with an example of how we can analyse athletic performance when we are knowledgeable in anatomy and the laws of mechanics. On the basis of such an analysis, strength and flexibility training should be designed to suit the different needs of the individual athlete. The sports technique we will examine here as an illustration, is the long jump (Figure 304). We will consider in sequence: (a) run-up, (b) take-off, (c) flight and (d) landing. For each of these we will determine which muscles are active and how they work (concentrically, eccentrically, statically). We will also determine which muscles require a wide range of movement, i.e. ought to be trained for flexibility.

Fig. 304.

A. Run-up

Look at the way your foot strikes the ground. To minimise the risk of injury, the outer border of your heel should be the part of your foot which first makes contact with the ground. Make sure your foot is pointing straight ahead of you, in the direction you are running. In general, striking the ground with the whole foot or the front of the foot first while the foot is pointed slightly outward, leads to complaints of pain in the lower leg (periostitis, p.61). Strong knee extensors (quadriceps femoris, p.49) are essential for swinging the leg forward in running but above all, for the correct heel strike. In the swing phase, the hip extensors (gluteus maximus and hamstrings) should be so long and so relaxed that they do not prevent a proper forward swing, i.e. hip flexion with pelvis pressed forward. During the forward swing the body is supported in the forefoot of the stance phase foot, therefore the calf muscles (triceps surae, p.57) must be strong enough to prevent the body sinking "backwards" at the ankle while the knee and hip extensors work. The calf muscles work statically during part of the stance phase, and concentrically in the latter stages (Figure 305).

The heel should be pulled towards the buttocks as early as possible, [i.e. concentric work for the knee flexors (hamstrings, p.53)], during the swing phase. This also provides flexibility training for the knee extensors (quadriceps, p.49) and keeps the moment of inertia of the leg to a minimum during its forward swing. Thus, the speed of the forward swing is dependent on the strength of the hip flexors and the knee flexors, as well as flexibility of the knee extensors. According to the above, training ought to concentrate on: 1, strengthening the iliopsoas, hamstrings, quadriceps and triceps surae, and 2, stretching the iliopsoas and quadriceps.

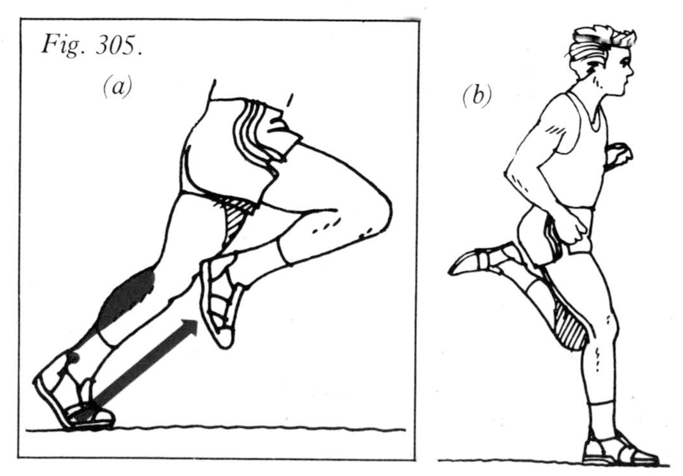
Fig. 305. (a) (b)

B. Take-off

At take-off the body has maximal speed. For a body travelling at a certain speed to reach as far as possible in a long jump, its angle of take-off should theoretically be about 45° (Figure 306). In practice, however, the jumper must hold back, i.e. steal some of his approach speed, in order to project himself upwards. To reach a take-off angle of 45°, the force with which he gathers for upward thrust must be so great and directed so far back that his speed after take-off would be too low. The difficulty of taking-off effectively lies thus in obtaining sufficient height without losing too much horizontal velocity.

Fig. 306.

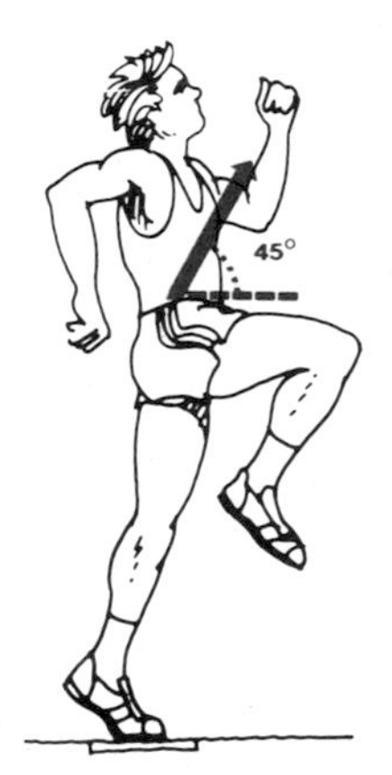

Theoretically, a jumper can take-off without reducing his speed only if the force of take-off is directed straight upwards. This would require a lightning swift upward thrust from a low position.

Fig. 307.

Short final stride.

In some take-off styles the jumper takes a long third "last stride" and in so doing lowers his centre of gravity. By making the final stride relatively short, he is able to plant his lift-off foot on the ground vertically from above (Figure 307). From that low position, there is even a possibility of projecting the body upwards *and* forwards.

Fig. 308.

Attempts to obtain height in this way — without losing speed — result in take-off angles of about 20–25° and — with a reduction in speed of about 30%. This style requires an explosive forward thrust of the hip, knee and ankle (Figure 308).

Fig. 309.

Long final stride.

The jumper may also obtain height by making a long final stride (Figure 309) and keeping his lift-off leg straighter than in the style outlined above. Now the muscles responsible for take-off are more hip extensor muscles (gluteal muscles, p.39) than knee extensors.

These styles of take-off in the long jump can be compared with the high jump styles. Dive Straddle (deep knee flexion and emphasised free arm and leg movements) and Fosbury Flop (greater speed of approach and bracing of the lift-off leg as strength of the hip is essential here).

We could also make comparisons between take-off styles in such sports as volleyball, football (to head the ball) and the like, when taking note of the speed of approach, i.e. the take-off styles requiring high approach speed compared with those requiring low approach speed.

C. Flight

Whatever the take-off style, the jumper usually has a certain forward rotation. This rotation can be "cancelled" by adopting certain in-the-air techniques such as [Figure 310 (a)] hang or (b) hitch-kick.

Fig. 310.

(a)

(a) If the jumper has a large forward rotation, he must lie outstretched in the air in order to avoid rotating so far that he arrives in an unfavourable position for landing. The speed of rotation is controlled by the body's moment of inertia (p.107). The more he folds together, the greater (faster) will be his rotation. The less rotation the jumper acquires from take-off, the sooner he can adopt a position that is advantageous for minimising air resistance.

(b)

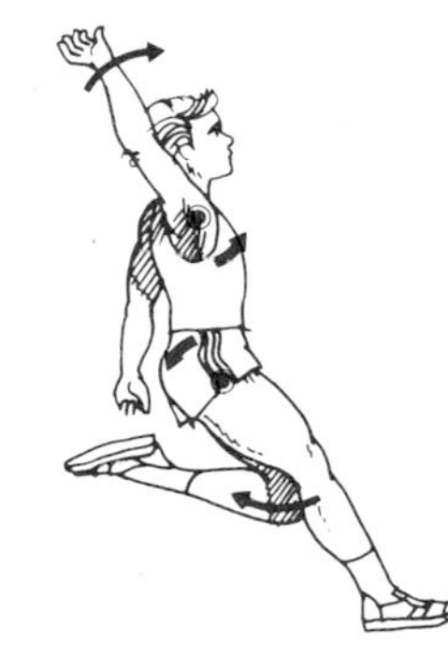

(b) If the rotation is "too great", it must be countered by an opposing movement, i.e. the arms and legs should rotate in the same direction. If the arms move according to the arrow ↗ the body will move in the opposite direction at the shoulder joint, i.e. forward rotation is *"cancelled"*. A backward rotation of the legs according to the arrow ↶ causes the pelvis to tip in the other direction at the hip joint, i.e. forward rotation is "cancelled".

D. Landing

Landing requires good flexibility of the hamstrings and of the lower back. A maximal hip piking before touchdown requires long muscles at the back of the thigh in order to avoid striking the ground with the heels too soon.

Maximal jump length is obtained when the heels touch down at the point where the centre of gravity of the body would have landed (Figure 311). If the heel strike is too far in front of this point, the jumper will land on his buttocks. If his heels touch down too close to the take-off board, he will be seen to rotate forward at touchdown and, of course, his result will be poor.

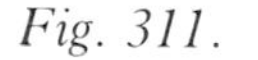

Fig. 311.

In a perfect landing, the jumper is pressed together like an accordion, which subjects his knees to great stress. Therefore, his training must include deep knee flexion in order to prevent acute injuries.

Chapter 7

HINTS FOR STRETCHING

Examples of strength training and stretching exercises for different muscle groups accompany each section of the book, each of which describes a part of the body. The following diagrams summarise and complement these examples with positions that are designed to lengthen muscles and thereby increase the range of movement of a given body part. Sometimes several different positions are given for the same muscle group. It is up to you to choose the position (positions) that is (are) best suited to your particular needs. If you are "stiff" certain positions will be suitable, or if your body is already rather flexible others should be chosen. Test the methods to find those that best suit you.

Some variations of stretching methods, namely, elastic stretching, stretching and the PNF-method, were described on pp.24–26. It should be pointed out that all types of flexibility training aim at producing a more flexible body; i.e. reaching the outer limits of the body's range of movement is the ultimate goal. Stretching is excellent for restoring a shortened muscle (due to injury or incorrect training methods) to its normal length. When a muscle has regained its normal length, it can be "taught" to function correctly by submitting it to flexibility training consisting of flexibility and elastic stretching exercises.

Dancers have trained flexibility with elastic stretching for centuries; they provide excellent proof that this method of training yields good results. It cannot be stressed too strongly that it is essential to warm up before undertaking intensive exercise. Although a person can raise the temperature of particular muscles by static contraction (which is a part of the PNF-method), if he wants to raise his entire body temperature substantially and at the same time subject his joints to all-round stress, he should begin any athletic activity with a well prepared "running–gymnastics-programme"

1. Stretching exercises should be part of warming-up. Do not be afraid to make easy elastic stretches towards the outer limits of your range of movement, after you have first stretched in the ordinary manner for about 30 seconds. Muscle spindles (p.21) are sensitive to both position (stretching) and speed (elastic stretching); therefore muscles must become accustomed to moving towards the body's outer limits at a moderate speed.
2. Either stretching exercises or PNF exercises should be a part of special training sessions where the aim is to increase the length of a particular muscle group. Once he has been given suitable instructions, the athlete ought to be able to conduct these training sessions himself during periods (three times a week to increase flexibility; once or twice a week to maintain it) that do not interfere with such other training as team practice.
3. Stretching should be practised after heavy training sessions in order to avoid stiffness, soreness and muscular aches the following day. In this situation, stretching has the effect of lightly massaging the muscles.

Keep the principal objective in mind: a more flexible body.

Positions for stretching and elastic stretching exercises for muscles which pass over:

The front of the ankle (extensors, p.60)

Fig. 312.

Do you find it difficult to sit comfortably on your heels? If so, this is due to your knees. Therefore, you must not try to lower yourself by stretching elastically if your knees hurt!

Goal: Stretched, or somewhat overstretched ankle.

Back part of the ankle (flexors, p.57)

Goal: That the muscles at the back of the ankle shall not restrict the natural flexibility of the ankle. Short foot muscles may produce high tension in the muscles at the front of the lower leg which could easily lead to "periostitis".

Fig. 313.

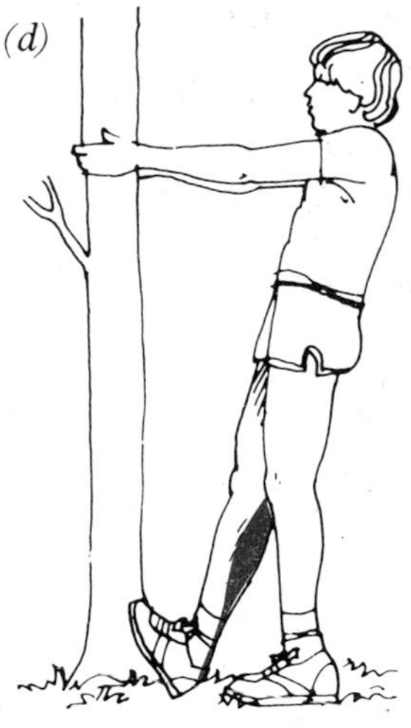

(d)

(c)

(a)

(b)

(e)

(f)

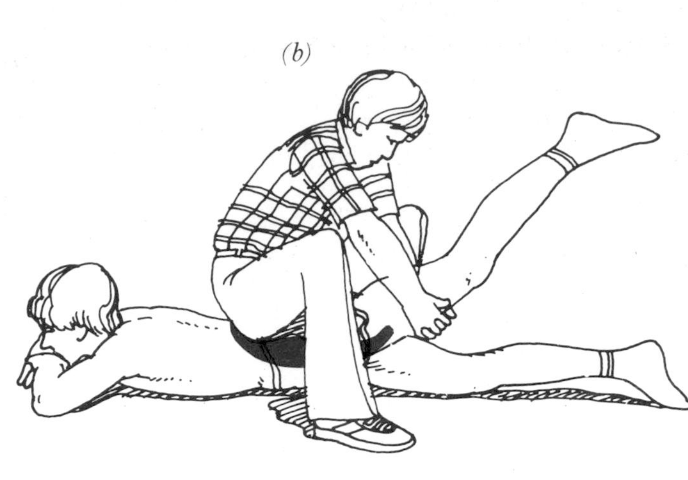

Knee joint and front part of the hip (quadriceps femoris and iliopsoas, pp.49 and 43)

Goal: To be capable of initiating a forward swing (during sprinting or kicking a football, for example) of a leg whose starting position is well behind the body.

Fig. 314.

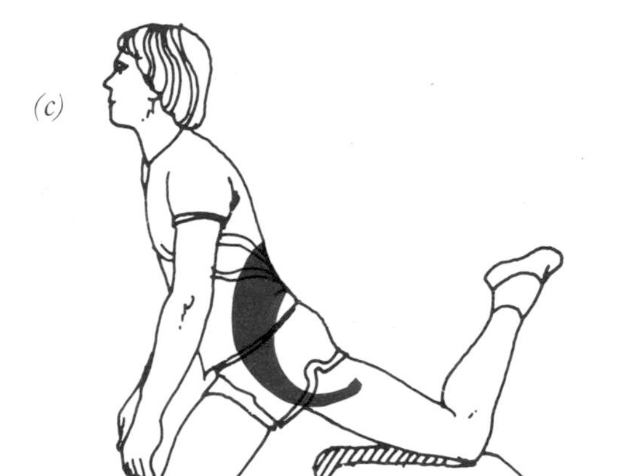

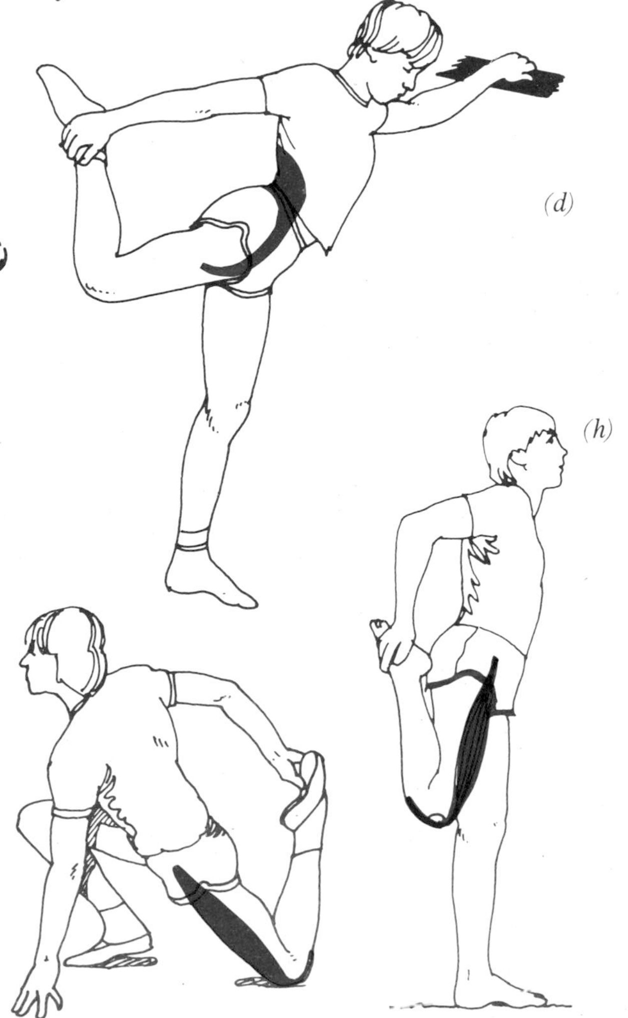

Knee joint and back part of the hip (hamstrings, p.53)

Goal: Sufficient flexibility to allow a person to "fall" forward at the hip and not tax his back unnecessarily by "bouncing". Short muscles at the back of the thigh generally lead to lower back pain.

Fig. 315.

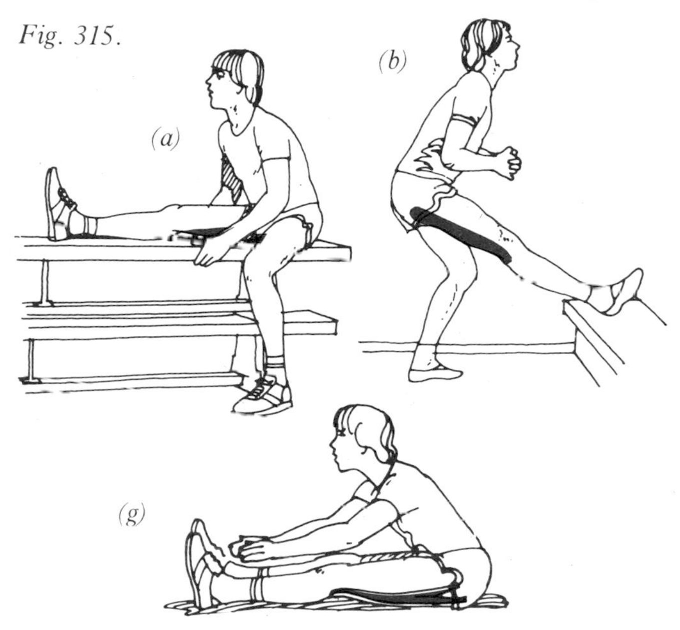

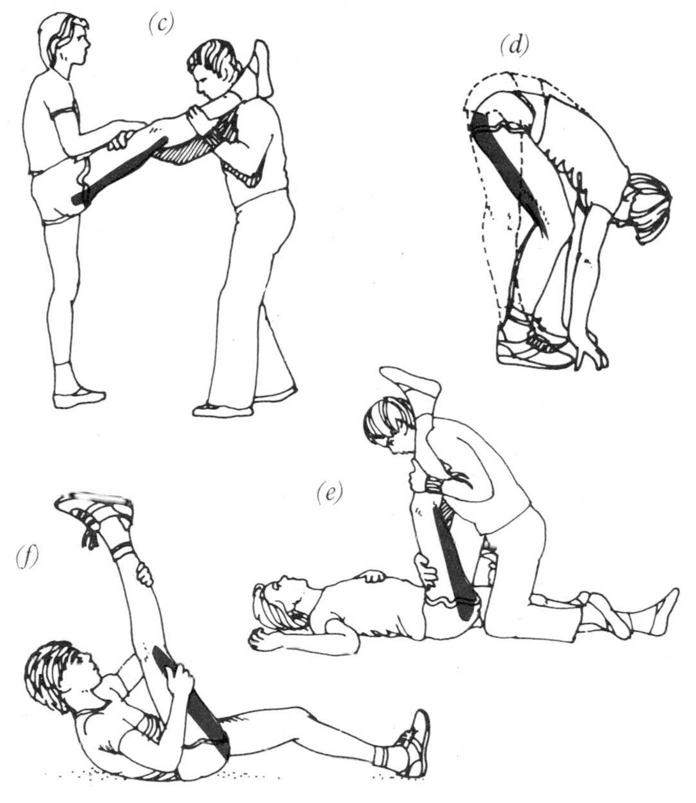

Fig. 316.

Outer part of the hip (abductors, p.39)

Goal: To increase flexibility of the hip and at the same time stretch the tendon band which extends from the crest of the hip bone down to the outer surface of the knee (iliotibial tract, Figure 90). A tendon band that is too taut can lead to friction injuries of the outer condyle of the shin bone (tibia) in, for example, the long distance runner (runner's knee).

Fig. 317.

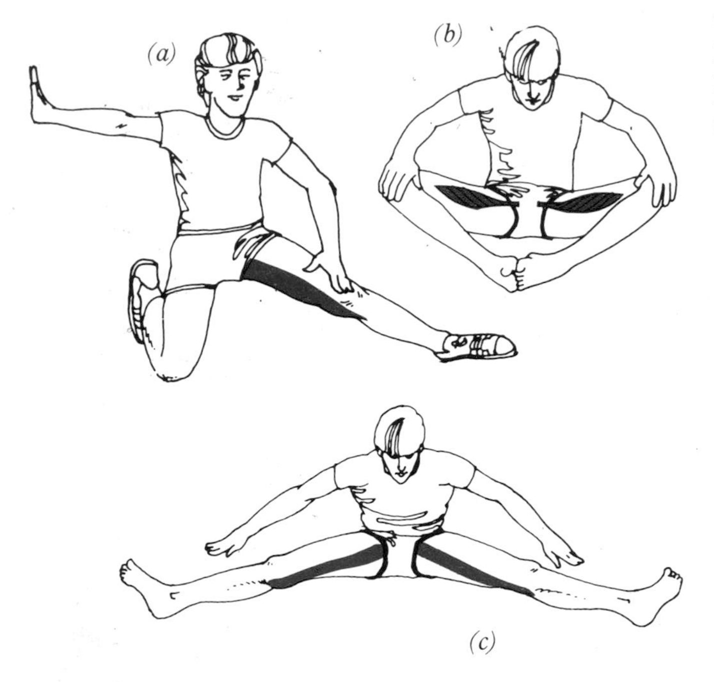

Inner part of the hip (adductors, pp. 41 and 42)

Goal: Increased flexibility for avoidance of groin injuries. Increased flexibility makes possible longer running strides, better forward swing in ice-skating, longer stopping strides when attacking in badminton or tennis, etc.

Back and abdomen (erector spinae and abdominal muscles, pp.68–73)

Goal: To increase flexibility of forward bending of the back. The diagrams (Figure 318) demonstrate positions that allow you to alternate between easy elastic stretching and relaxation.

(a) (b)

Fig. 318.

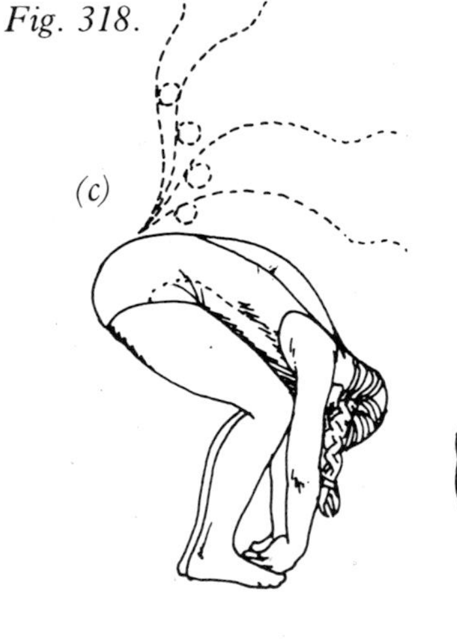

(d)

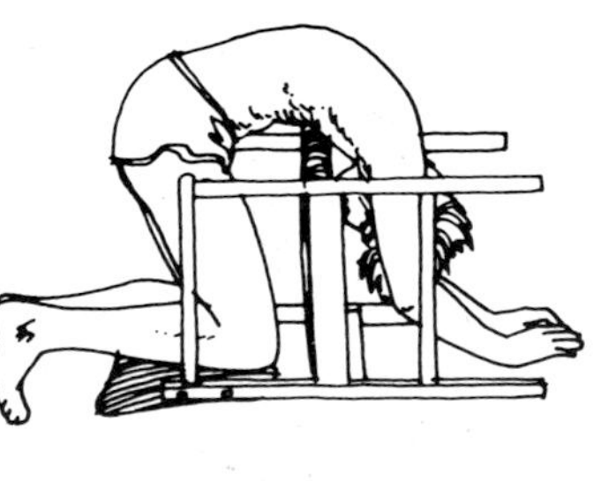

(e)

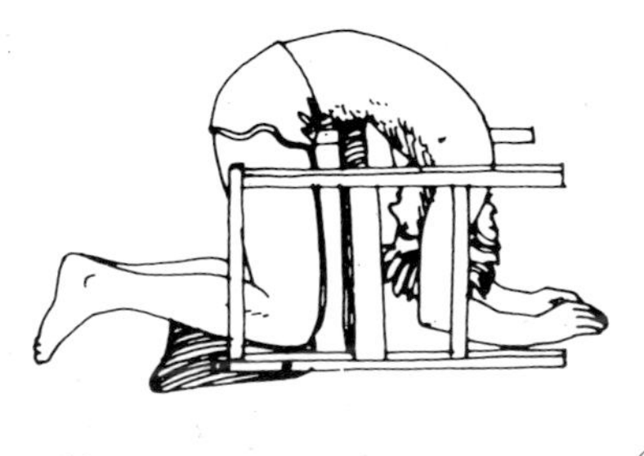

(f)

(g)

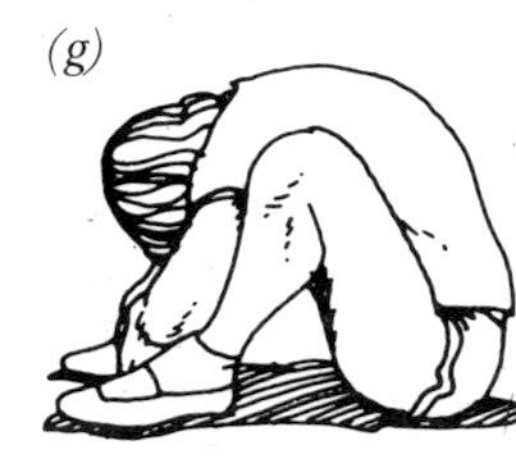

(h)

(i)

The diagrams shown in Figure 319 demonstrate elastic stretching and positions suitable for sideward bending and rotation of the back.

Fig. 319.

Shoulder joint (pp.78–86)

Goal: Increased flexibility of the shoulder which is often necessary in such sports as badminton, tennis, golf, handball and swimming.

Fig. 320.

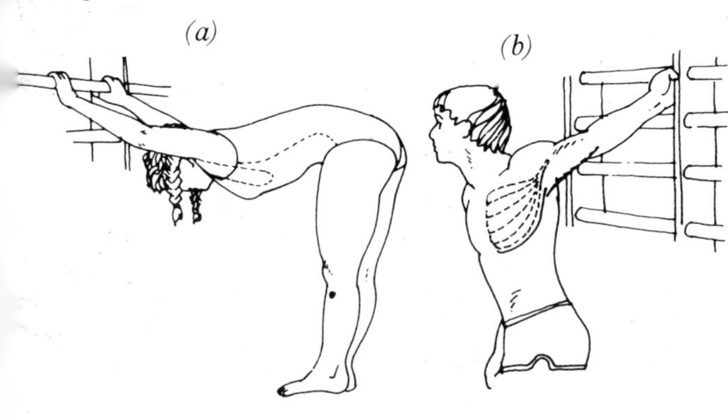

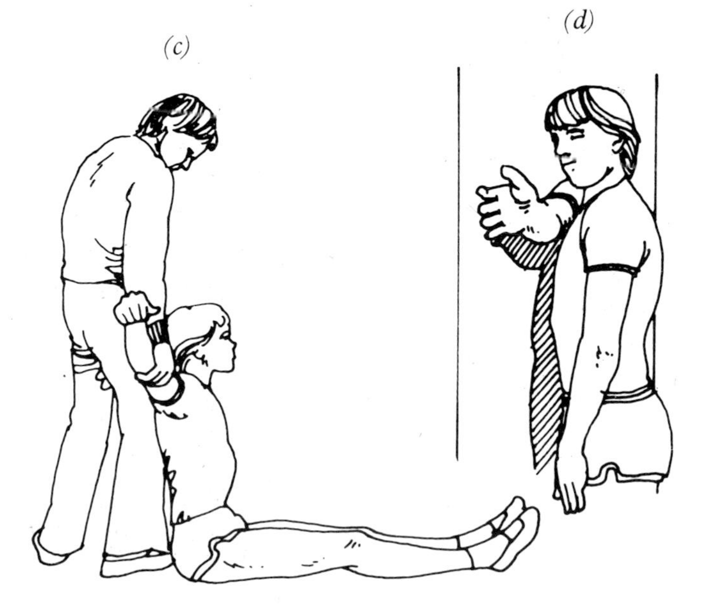

Fig. 320.

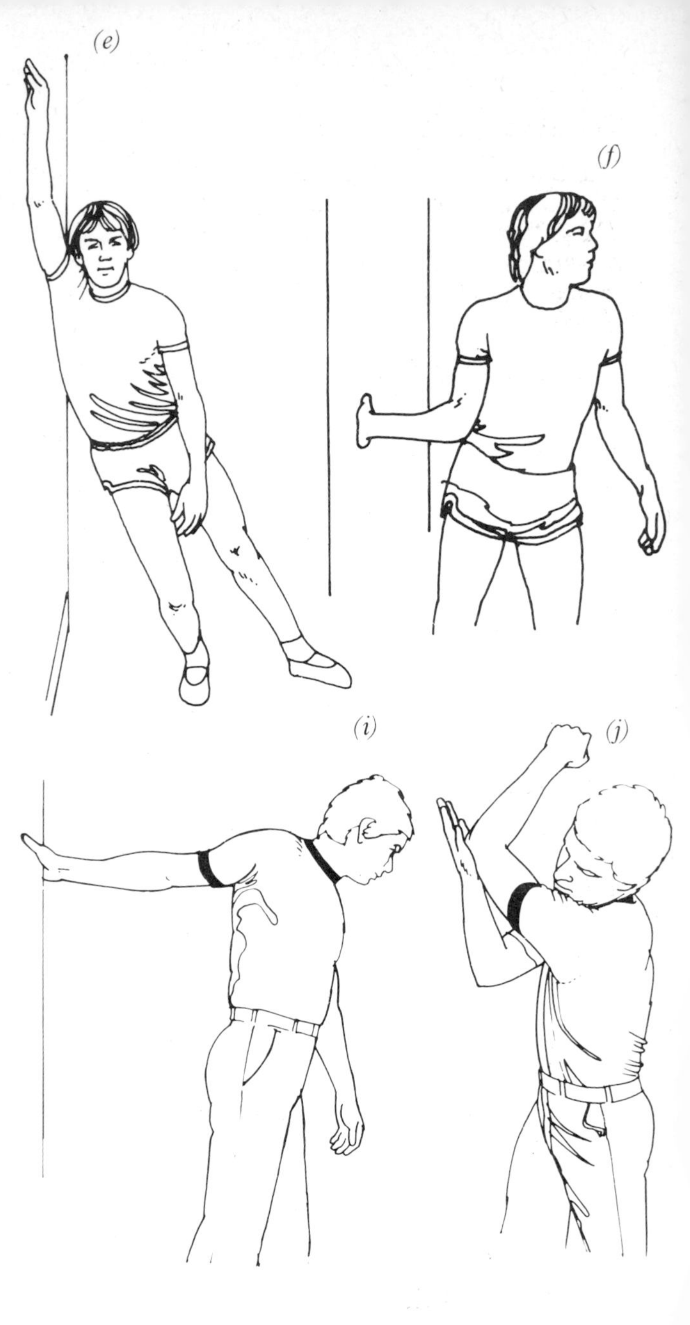

Wrist (p.91)

Goal: Dorsiflexion flexibility. Applies to racket sports.

Fig. 321.

Towel routine

It is easy to train strength and flexibility with the aid of a towel. The four exercises shown in Figures 322–325 provide an example of how you can work dynamically with elastic stretches in order to increase the strength and flexibility of your shoulders.

The position in exercise 3 is good for stretching the triceps brachii. There is seldom any need of stretching exercises for the muscles which pass over the elbow joint.

By using your imagination, you can easily develop your own exercises. It is a good idea to work in pairs. In that way you can both provide sufficient resistance to your partner in exercises that demand strength.

Fig. 322.

Exercise 1. keep the towel taut and "dry" the back of your neck. Make sure that your grip is such that you can pull one of your arms with the other and thereby increase its flexibility "a little bit" each time. Do 20–30 of these.

Fig. 323.

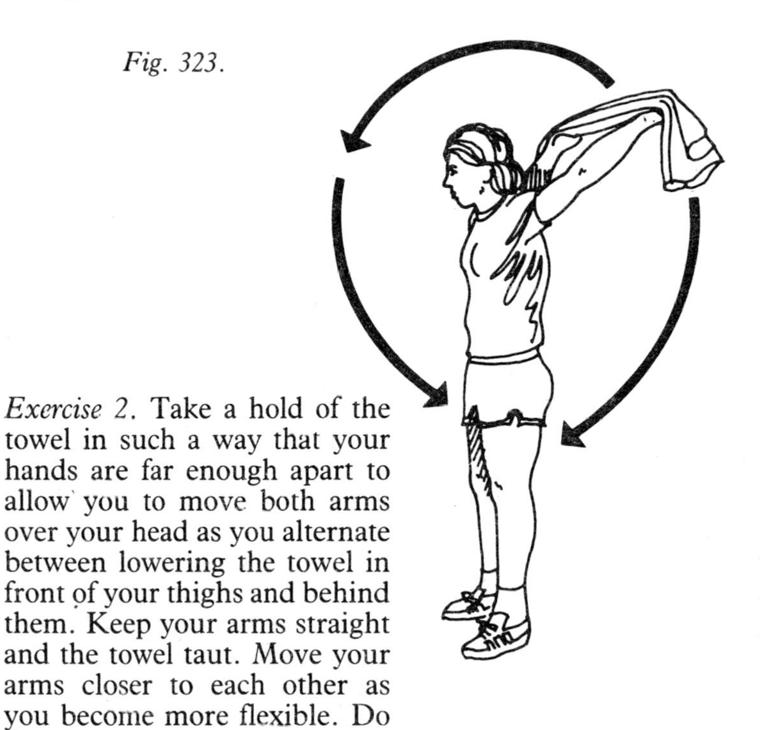

Exercise 2. Take a hold of the towel in such a way that your hands are far enough apart to allow you to move both arms over your head as you alternate between lowering the towel in front of your thighs and behind them. Keep your arms straight and the towel taut. Move your arms closer to each other as you become more flexible. Do the exercise 20–30 times.

Fig. 324.

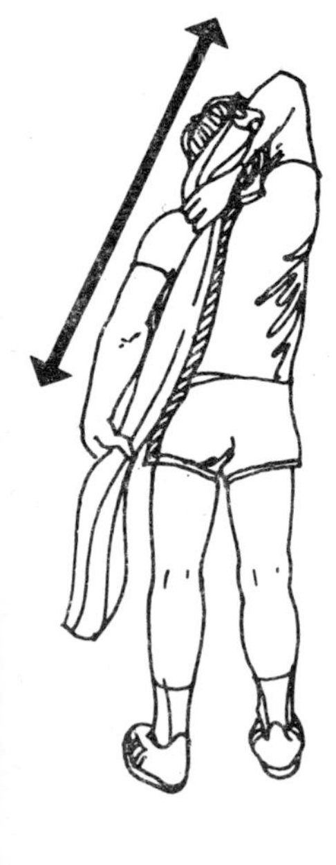

Exercise 3. "Dry your back" Pull your right (bent) arm down with the aid of your left (straight) arm. After that, pull your straight arm up "a little bit further" with the aid of your bent arm. Do the exercise 10 times with bent right arm and 10 times with bent left arm.

Fig. 325.

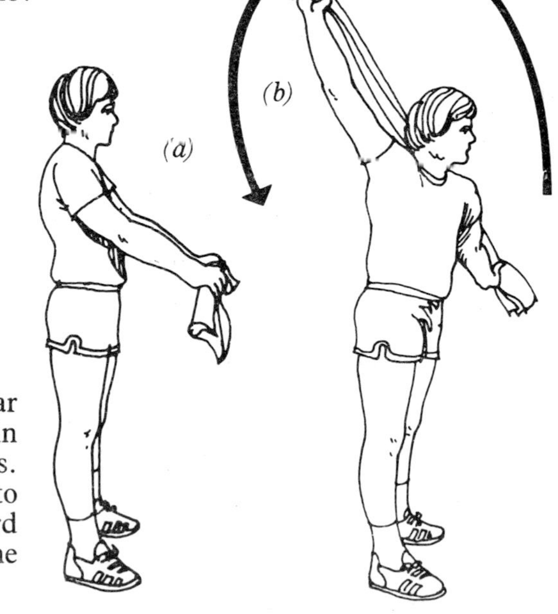

Exercise 4. (a) Hold the towel with your hands so far apart that you can (b) move the towel alternately in front of and behind your body with straight arms. Swimming with "backstrokes" takes the towel to the behind-your-back position, and "forward strokes" takes it to its original position. Do the exercise 20 times for each arm.

Chapter 8

REVIEW OF RELEVANT MUSCLES

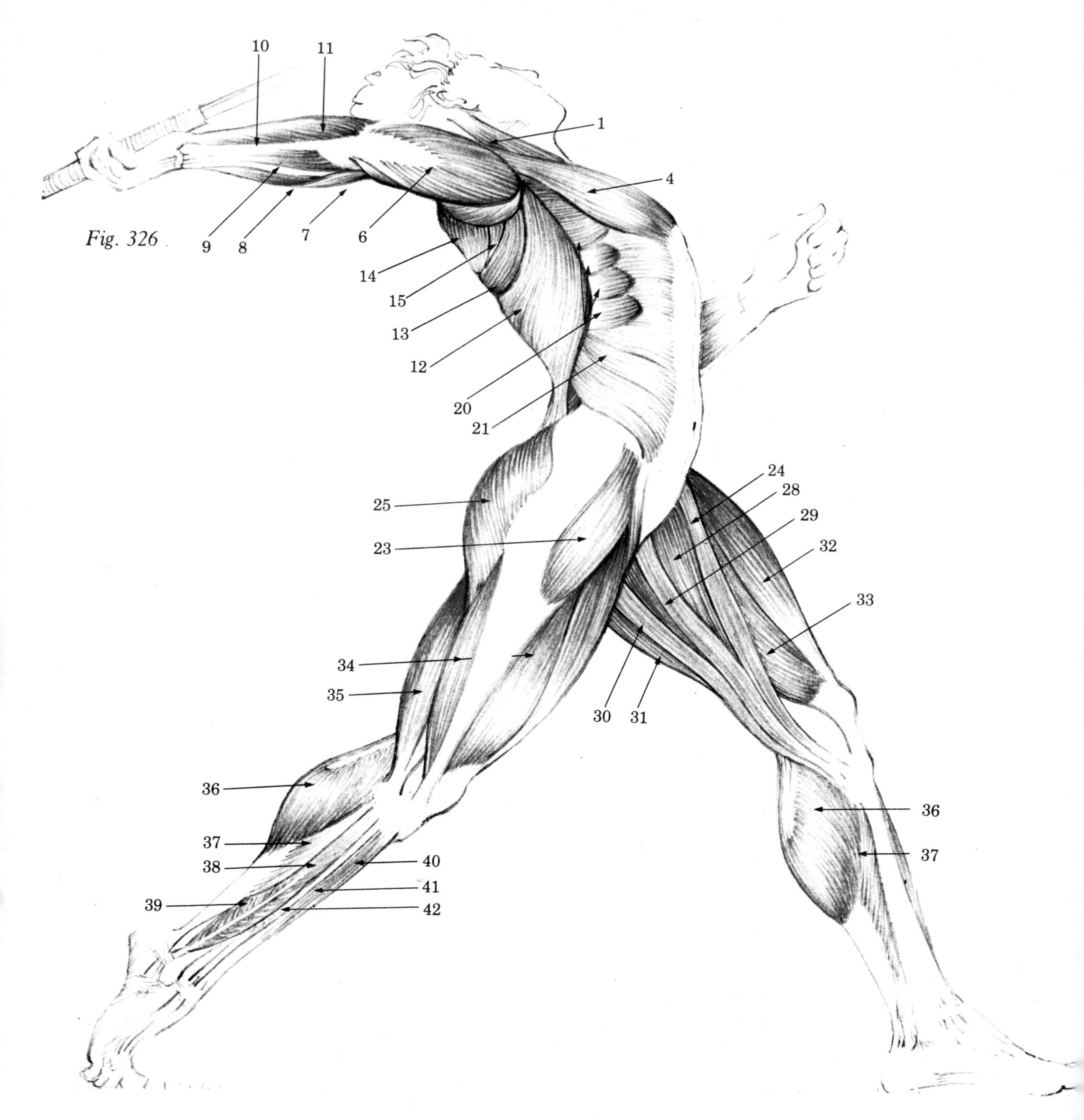

Fig. 326

LIST OF NAMES OF THE MUSCLES SHOWN IN FIGURES 326, 327 and 328

1 Sterno-cleido-mastoid
2 Trapezius
3 Deltoid
4 Pectoralis major
5 Biceps brachii
6 Triceps brachii
7 Brachioradialis
8 Extensor carpi radialis longus and brevis
9 Extensor digitorum
10 Extensor carpi ulnaris
11 Flexor carpi ulnaris
12 Latissimus dorsi
13 Teres major
14 Infraspinatus
15 Teres minor
16 Supraspinatus
17 Rhomboid major
18 Rhomboid minor
19 Levator scapulae
20 Serratus anterior
21 External oblique
22 Rectus abdominis
23 Tensor fascia latae
24 Sartorius
25 Gluteus maximus
26 Iliopsoas
27 Pectineus
28 Adductor longus
29 Gracilis
30 Semitendinosus
31 Semimembranosus
32 Rectus femoris
33 Vastus medialis
34 Vastus lateralis
35 Biceps femoris
36 Gastrocnemius
37 Soleus
38 Peroneus longus
39 Peroneus brevis
40 Tibialis anterior
41 Extensor hallucis longus
42 Extensor digitorum longus
43 Tibialis posterior
44 Flexor hallucis longus
45 Flexor digitorum longus

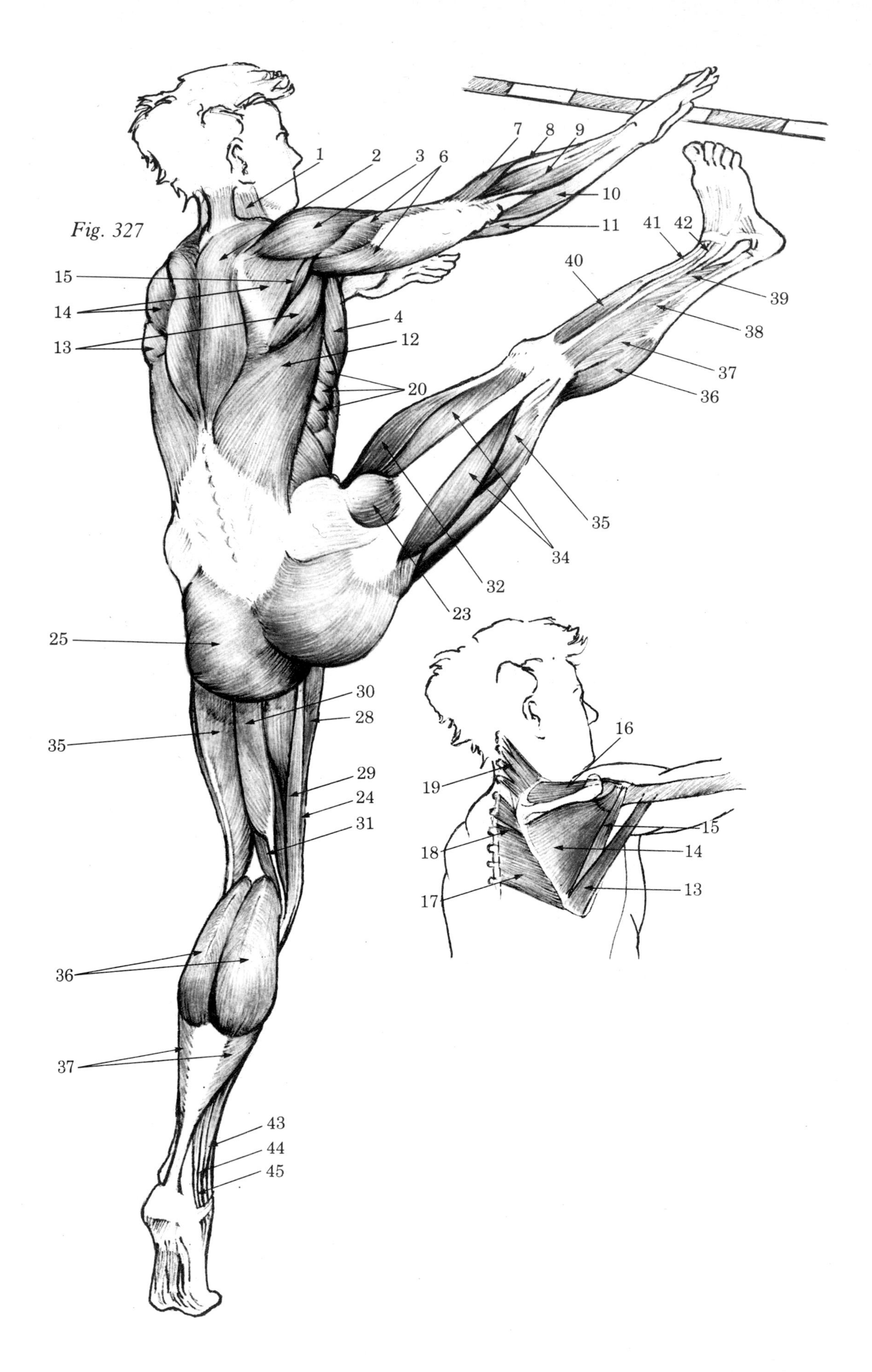
Fig. 327
1
2
3
6
7
8
9
10
11
41
42
40
39
38
37
36
15
14
13
4
12
20
35
34
32
23
25
30
28
35
29
24
31
36
37
43
44
45
16
19
15
18
14
17
13

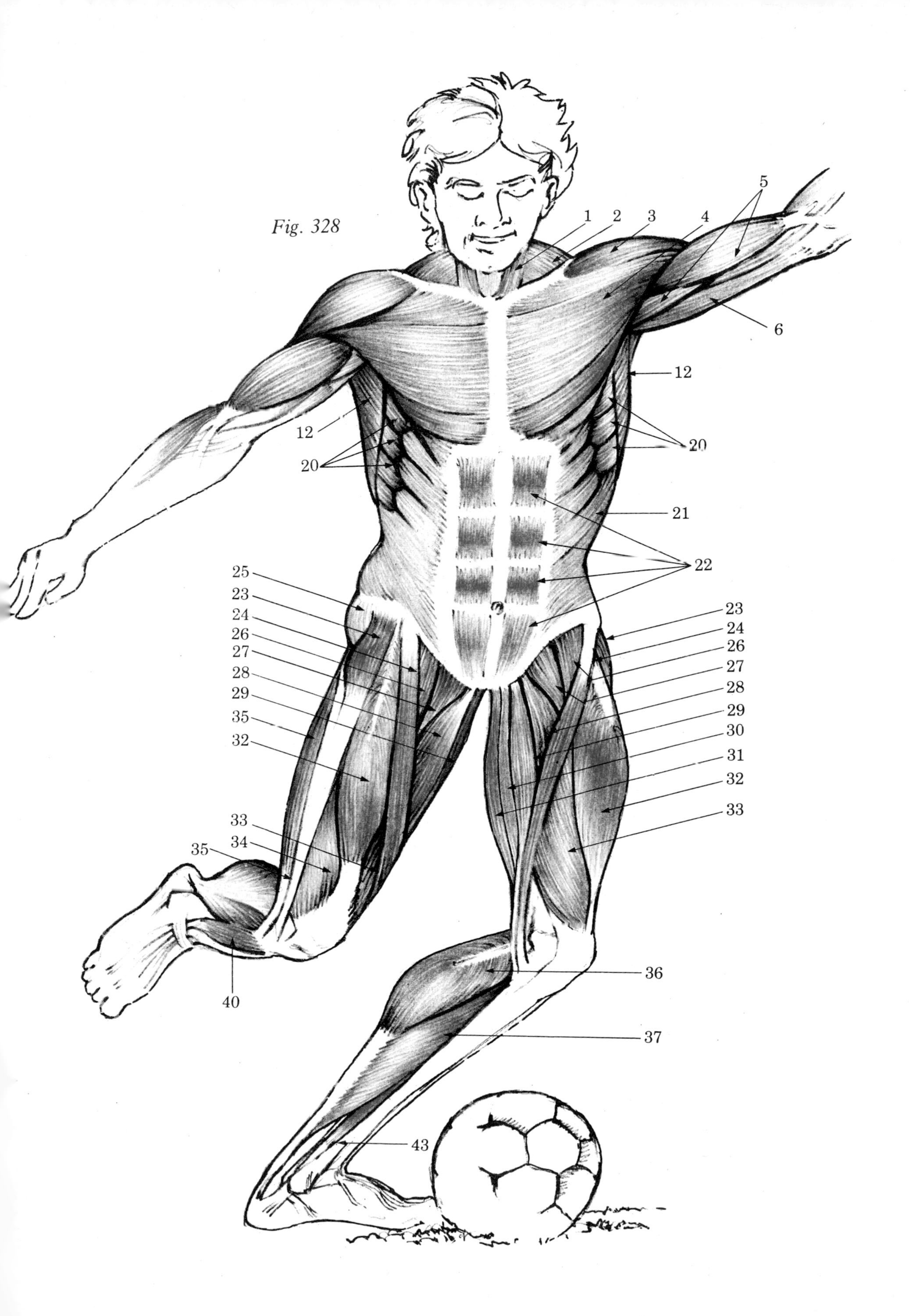
Fig. 328
1
2
3
4
5
6
12
12
20
20
21
22
25
23
24
26
27
28
29
35
32
33
34
35
23
24
26
27
28
29
30
31
32
33
40
36
37
43

Tables of the body's most important muscles, together with their origin, insertion and function. In the accompanying diagrams: the origin and insertion are marked.

Muscles which pass across the hip joint only (pp.39–46)

Muscle	Origin	Insertion	Function
Gluteus maximus (large buttock muscle)	Posterior part of the hip bone crest, the sacrum and coccyx and coccyx	Outer surface of the femur just below the greater trochanter and a strong tendon band on its outer side (iliotibial tract)	Straightens and adducts the hip, rotates the thigh outwards, and takes part in straightening the knee
Gluteus medius (intermediate buttock muscle)	Outer surface of the ilium	Greater trochanter	Primarily an abductor and rotator of the hip
Gluteus minimus (small buttock muscle)	Outer surface of the ilium immediately beneath and behind the intermediate buttock muscle	Greater trochanter	Same as the intermediate buttock muscle. Both are active during walking, and both stabilise the pelvis
Pectineus (comb muscle)	Upper border of the pubic bone	High on the posterior surface of the femur (pectineal line)	Adducts, flexes and rotates the hip outwards
Adductor brevis (short adductor)	Lower border of the pubic bone	Linea aspera	Adducts the hip and rotates it outwards
Adductor longus (long adductor)	Near the symphysis pubis of the pubic bone	Linea aspera	Adducts the hip
Adductor magnus (large adductor)	Two parts. One from the pubic bone and the other from the ischial tuberosity	Linea aspera and the medial condyle	Adducts the hip. Can also rotate it inwards
Psoas major (great lumbar muscle)	Side of dorsal vertebra 12 and lumbar vertebrae 1–5	These two muscles go under the common name of iliopsoas, which is attached to the lesser trochanter	Bends the hip. Rotates the leg outwards and can also bend the vertebral column sidewards
Iliacus (haunch muscle)	Entire inner surface of the ilium		

Muscles which pass across both the hip and knee joints (pp.49–56)

Muscle	Origin	Insertion	Function
Rectus femoris (straight thigh muscle)	Anterior inferior iliac spine and margin of the acetabulum	Kneecap (patella) via quadriceps tendon	Straightens the knee and bends the hip
Gracilis (slender thigh muscle)	Pubic bone	As for semitendinosus	Adducts the hip and bends the knee and rotates it inwards
Tensor fascia latae (tensor of the fascia of the thigh)	Outer surface of the anterior superior iliac spine	Into strong tendon band of the thigh fascia (iliotibial tract)	Bends and abducts the hip and straightens the knee
Biceps femoris (two-headed thigh muscle)	Ischial tuberosity and linea aspera	Head of the fibula	Straightens (extends) the hip. Bends the knee and rotates it outwards
Semitendinosus	Ischial tuberosity	Top of the inner surface of the tibia	Straightens the hip. Bends the knee and rotates it inwards
Semimembranosus	Ischial tuberosity	Several sites on and around the internal condyle of the tibia	Straightens the hip. Bends the knee and rotates it inwards
Sartorius (tailor's muscle)	Anterior superior iliac spine	As for semitendinosus	Bends, abducts and rotates the hip outwards. Bends the knee and rotates it inwards

Muscles which pass across the knee joint only (p.50)

Muscle	Origin	Insertion	Function
Vastus medialis	Medial and posterior surface of the femur	Direct to the kneecap	Together with the straight thigh muscle, these muscles form the so-called four-headed thigh muscle (quadriceps femoris) The vasti muscles straighten the knee. The straight thigh muscle also bends the hip
Vastus intermedius	Anterior surface of the femur	Kneecap via the quadriceps tendon	
Vastus lateralis	Posterior surface of the femur	Kneecap via the quadriceps tendon	
Popliteus	Posterior surface of the lateral femoral condyle	Posterior surface of the internal condyle of the tibia	Bends the knee and rotates it inwards. ("Unlocks" the knee joint)

Muscles which pass across both the knee joint and the ankle (p.57)

Muscle	Origin	Insertion	Function
Gastrocnemius (twin calf muscle)	Posterior surface of the medial and lateral femoral condyles	Tuber calcanei (heel bone)	Bends the knee and straightens the ankle (plantarflexion)

Muscles which pass across the ankle only (p.60)

Muscle	Origin	Insertion	Function
Soleus (flounder muscle)	Head of the fibula and from the oblique line of the tibia	Heel bone. Joins with the twin calf muscle to form the three-headed calf muscle (triceps surae)	Standing on your toes (plantarflexion)
Tibialis anterior (anterior shin bone muscle)	Upper end of the tibia	Intermediate cuneiform and the 1st metatarsal bone	Dorsiflexion and supination
Extensor hallucis longus (long great toe extensor)	Fibula and the membrane situated between the fibula and tibia	Great toe	Dorsiflexion and pronation of the ankle. Straightens the great toe
Extensor digitorum longus (long toe extensor)	Membrane between the fibula and tibia as well as the fascia of the lower legs	All the toes except the great toe	Dorsiflexion and pronation of the ankle. Stretches the toes
Peroneus longus (long calf muscle)	Upper part of the fibula	Its tendon of insertion passes behind the external malleolus, crosses the sole of the foot and is inserted into the great toe and the internal cuneiform bone	Builds up the transverse arch. Dorsiflexion and pronation of the ankle
Peroneus brevis (short calf muscle)	Lower part of the fibula	Fifth metatarsal bone	Dorsiflexion and pronation of the ankle
Flexor hallucis longus (long flexor of the great toe)	Posterior surface of the fibula	Underside of the great toe	Plantarflexion and supination. Bends the toes
Flexor digitorum longus (long toe flexor)	Posterior surface of the tibia	Underside of the toes (excepting the great toe)	Plantarflexion and supination. Bends the great toe
Tibialis posterior (posterior shin bone muscle)	Posterior surface of the fibula and tibia	Underside of the naviculare	Plantarflexion and supination

Fig. 329.
O = (origin) Dark red
I = (insertion) Pink

External view of the right hip bone

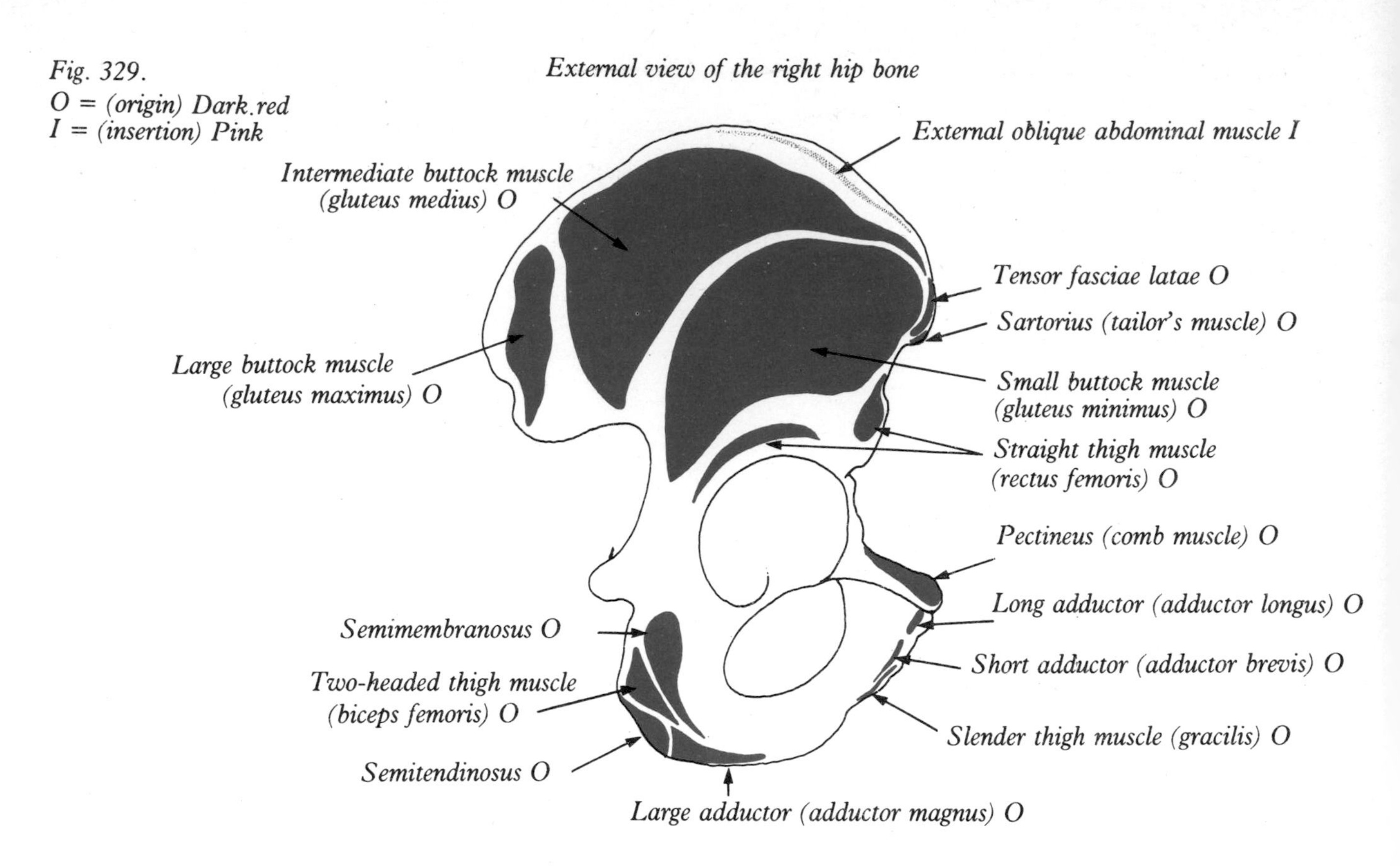

Fig. 330. (a) Right thigh bone viewed from the front. (b) Right thigh bone viewed from the back.

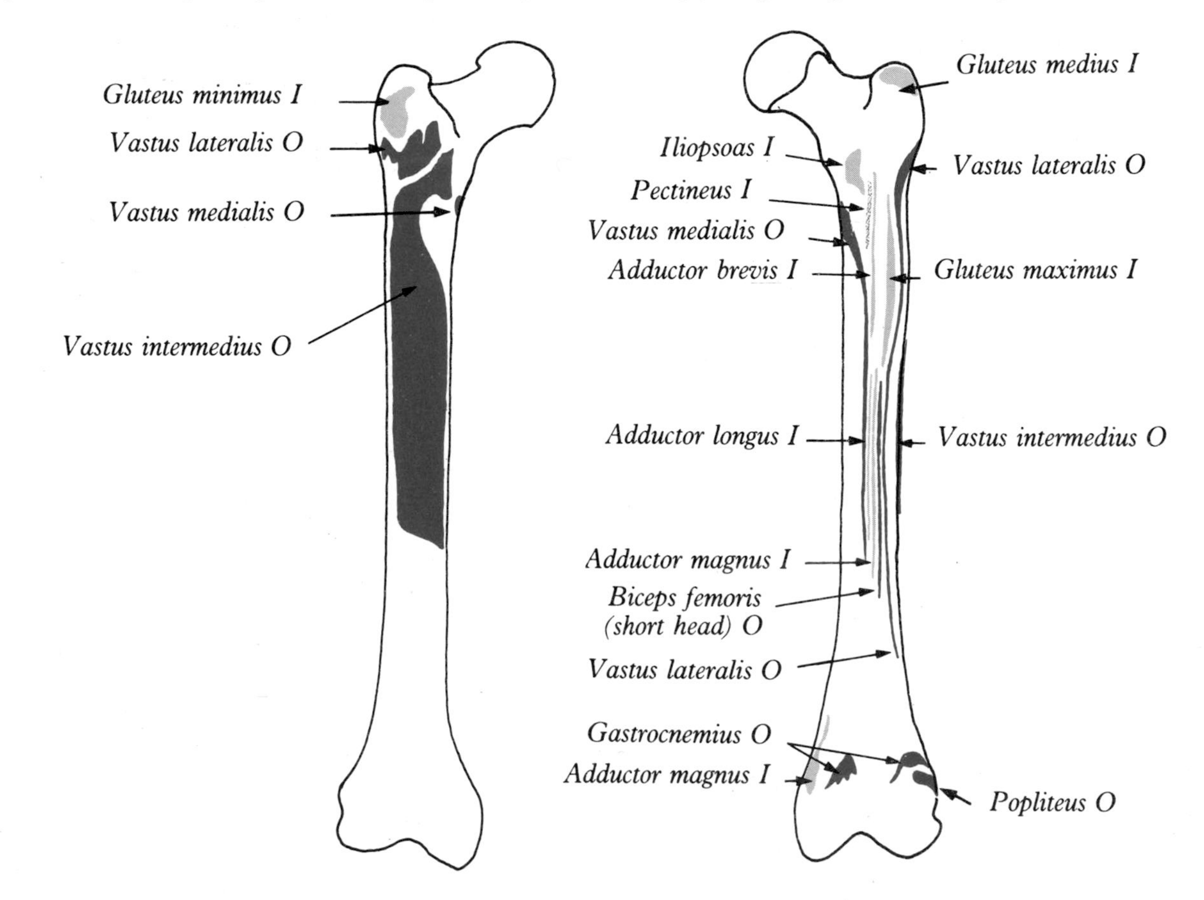

Fig. 331. Internal view of the right hip bone.

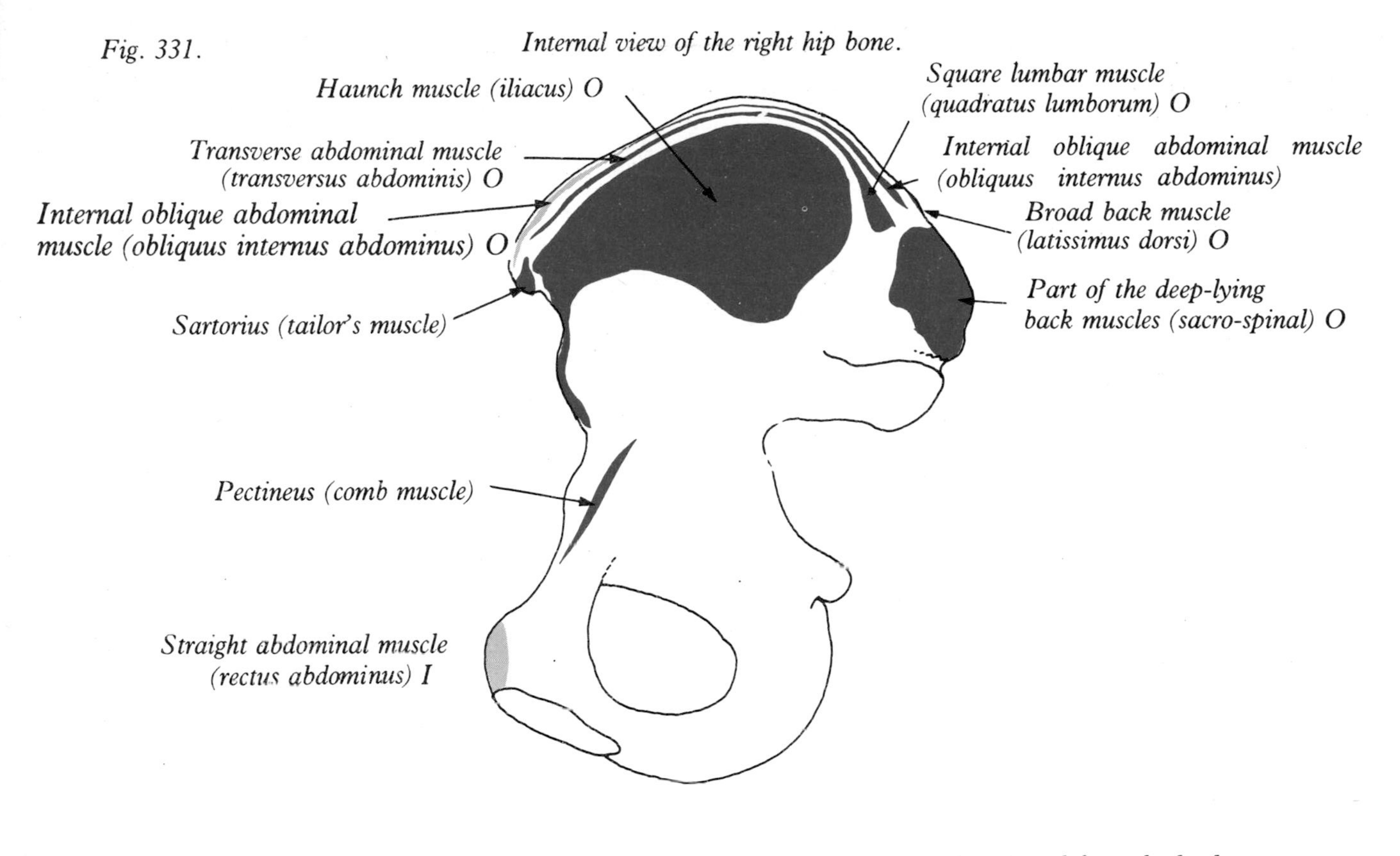

Fig. 332. (a) Right lower leg viewed from the front.

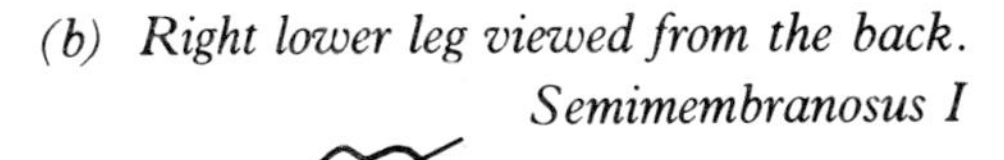
(b) Right lower leg viewed from the back.

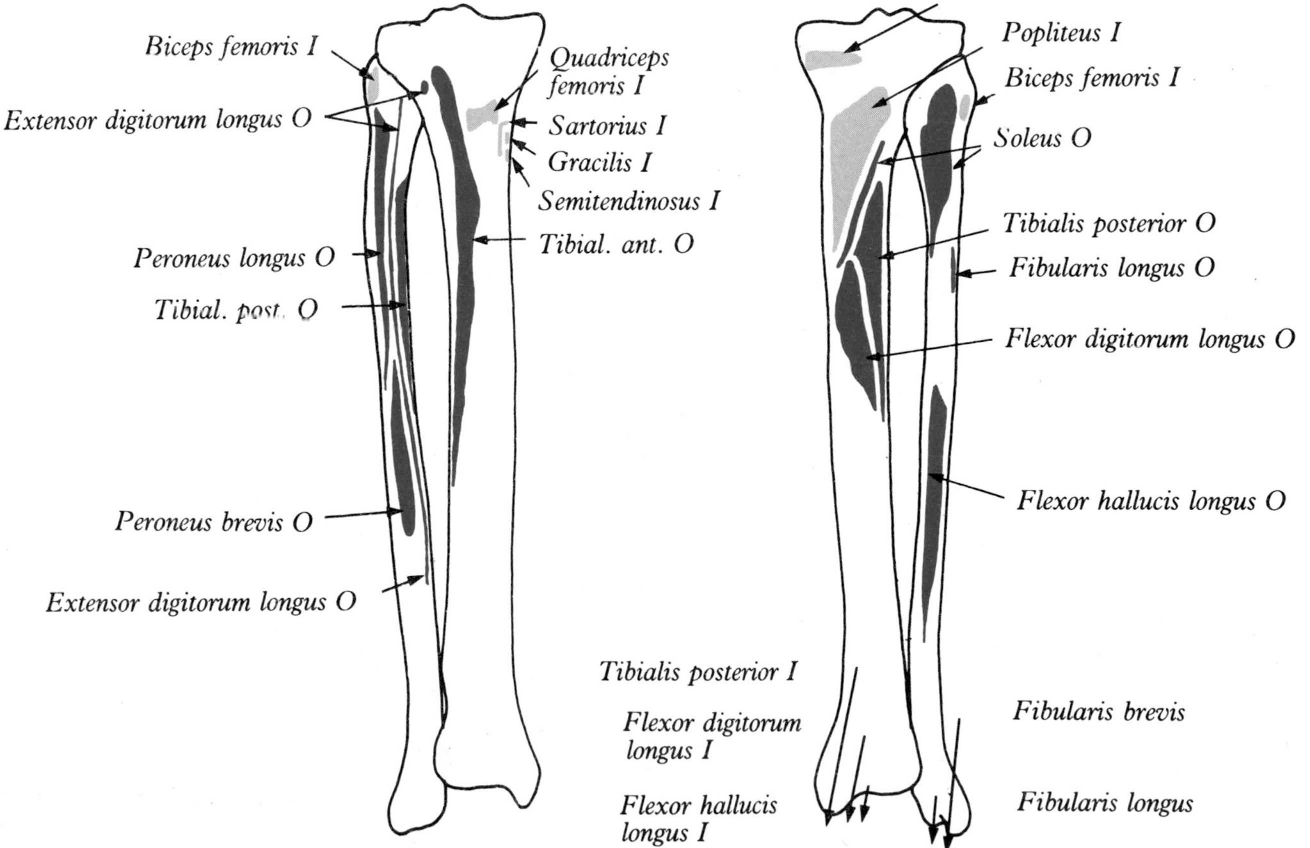

Muscles whose origin is on the shoulder blade and insertion on the upper arm (pp.80 and 81)

Muscle	Origin	Insertion	Function
Supraspinatus (above-the-shoulder-blade-spine muscle)	Area above the shoulder blade spine (fossa supraspinata)	Greater external tuberosity of the upper arm bone (greater tuberosity)	Adducts and rotates the arm outwards
Teres major (greater round muscle)	Inferior angle of the shoulder blade (angulus inferior)	Anterior surface of the upper arm bone (medial lip of biceps groove)	Adducts the arm and rotates it inwards. (Assists the broad back muscle)
Infraspinatus (below-the-shoulder-blade-spine muscle)	A large part of the surface underneath the shoulder blade spine (fossa infraspinata)	Greater external tuberosity (greater tuberosity)	Rotates the arm outwards
Teres minor (lesser round muscle)	Outer border of the shoulder blade (margo lateralis)	Greater external tuberosity (greater tuberosity)	Rotates the arm outwards
Subscapularis (anterior shoulder blade muscle)	Entire inner surface of the shoulder blade (fossa subscapularis)	Lesser anterior tuberosity (lesser tuberosity)	Rotates the arm inwards
Coracobrachialis	Crow's beak projection	Inner surface of the upper arm	Swings the arm forward

Muscles whose origin is on the trunk and insertion on the shoulder blade (pp.81 and 82)

Muscle	Origin	Insertion	Function
Levator scapulae	Cervical vertebrae 1–4	Superior angle of the shoulder blade	Raises the shoulder blade
Rhomboideus	Cervical vertebrae 6 and 7, dorsal vertebrae 1–4	Inner border of the shoulder blade	Adducts and rotates the shoulder blade inwards
Trapezius	Base of the skull. Cervical and dorsal vertebrae	Shoulder blade spine and external part of the collar bone	Adducts and rotates the shoulder blade outwards. Turns the head bends the neck backwards
Serratus anterior (anterior serrated muscle)	Ribs 1–9	Inner border of the shoulder blade	Stabilises the shoulder blade when the hand presses against an object
Pectoralis minor (lesser chest muscle)	Ribs 3–5	Crow's beak projection (coracoid process)	Lowers the shoulder blade

Muscles whose origin is on the trunk and insertion on the arm (pp.82–84)

Muscle	Origin	Insertion	Function
Pectoralis major (greater chest muscle)	Inner part of the collar bone, breastbone and part of the costal cartilage	Greater external tuberosity (greater tuberosity and lateral lip of biceps groove)	Adducts the arm and rotates it inwards. Pulls the arm in front of the chest from any position
Deltoideus (deltoid muscle)	Outer part of the collar bone and the shoulder blade spine (spina scapulae)	Along the shaft of the upper arm (deltoid tuberosity)	Takes part in all movements of the upper arm
Latissimus dorsi (broad back muscle)	Dorsal vertebrae 6–12. Posterior part of the hip bone crest and ribs 9–12. Rump bone	Below the lesser anterior tubercle of the upper arm (floor of the biceps groove)	Swings the arm backwards and rotates it inwards

Muscles which pass across both the shoulder and elbow joints (pp.87 and 89)

Muscle	Origin	Insertion	Function
Biceps brachii (two-headed arm muscle)	(1) Coracoid process (2) just above the articular surface of the shoulder blade (supraglenoid tubercle)	Anterior surface of the radius (tuberositas radii). Is also connected to the ulna by way of a tendon band	Bends and supinates the elbow. Swings the shoulder joint forwards
Triceps brachii (three-headed arm muscle)	(1) Below the articular surface of the shoulder blade (infraglenoid tubercle), (2) posterior surface of the upper arm, (3) posterior surface of the upper arm	Elbow outgrowth (olecranon) and joint capsule	Stretches the elbow and protects the shoulder joint by keeping the capsular ligament taut

Muscles which pass across the elbow joint only (pp.87 and 91)

Muscle	Origin	Insertion	Function
Brachialis (upper arm muscle)	Greater part of the anterior surface of the upper arm	Coronoid process	Bends the elbow
Brachioradialis (arm-radius muscle)	External condyle of the upper arm (lateral epicondyle)	At the styloid process of the radius (styloid process)	Bends the elbow. Pronation or supination depending on the start position
Anconeus	Posterior surface of the upper arm	Posterior surface of the ulna	Straightens the elbow
Supinator	External condyle of the upper arm	Outer surface of the radius	Supinates the forearm
Pronator teres	Internal condyle of the upper arm	Outer surface of the radius	Pronates the forearm

Muscles which affect both the forearm and wrist (p.92)

Muscle	Origin	Insertion	Function
Extensor digitorum (finger extensor)	External condyle of the upper arm (lateral epicondyle)	Posterior surface of the finger bones (excepting the thumb)	Straightens the fingers, the wrist and finally the elbow
Extensor carpi radialis longus and brevis (long and short radial-wrist extensors)	At the external condyle of the upper arm	Posterior of the 2nd and 3rd metacarpals	Associated with extension and abduction of the wrist joint
Extensor carpi ulnaris (ulna-wrist extensor)	External condyle of the upper arm	Posterior surface of the 5th metacarpal bone	Associated with extension and adduction of the wrist
Flexor digitorum superficialis (superficial finger flexor)	Internal condyle of the upper arm and anterior surface of the radius	Middle row of the phalanges	Bends the elbow and fingers. Assists in flexing the wrist
Flexor carpi radialis (radial-wrist flexor)	Internal condyle of the upper arm	Anterior surface of the 2nd and 3rd metacarpals	Bends the elbow. When acting alone it flexes the wrist. Can also assist in pronating the forearm and hand
Flexor carpi ulnaris (ulna-wrist flexor)	Internal condyle and the inner margin of the elbow outgrowth (olecranon)	Pisiform, hamate bone and 5th metacarpal bone	When acting alone it flexes the wrist and, by continuing to contract, it bends the elbow

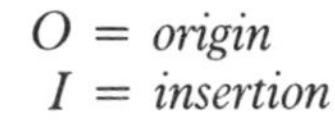

Fig. 333. *Right shoulder blade viewed from the back.*

O = origin
I = insertion

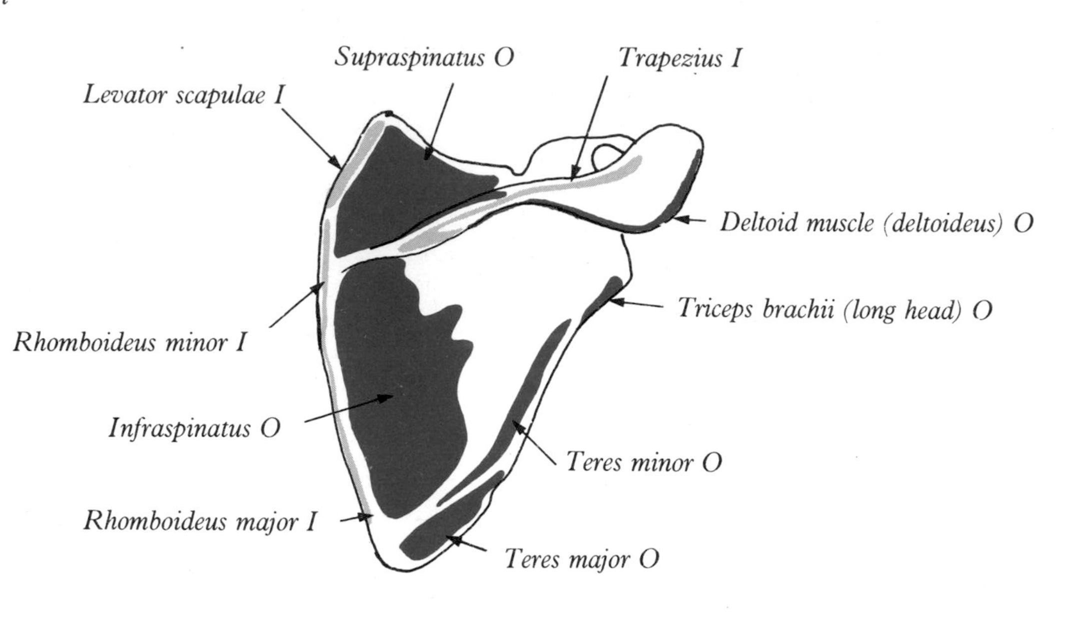

Fig. 334. *(a) Right upper arm viewed from the front.* *(b) Right upper arm viewed from the back.*

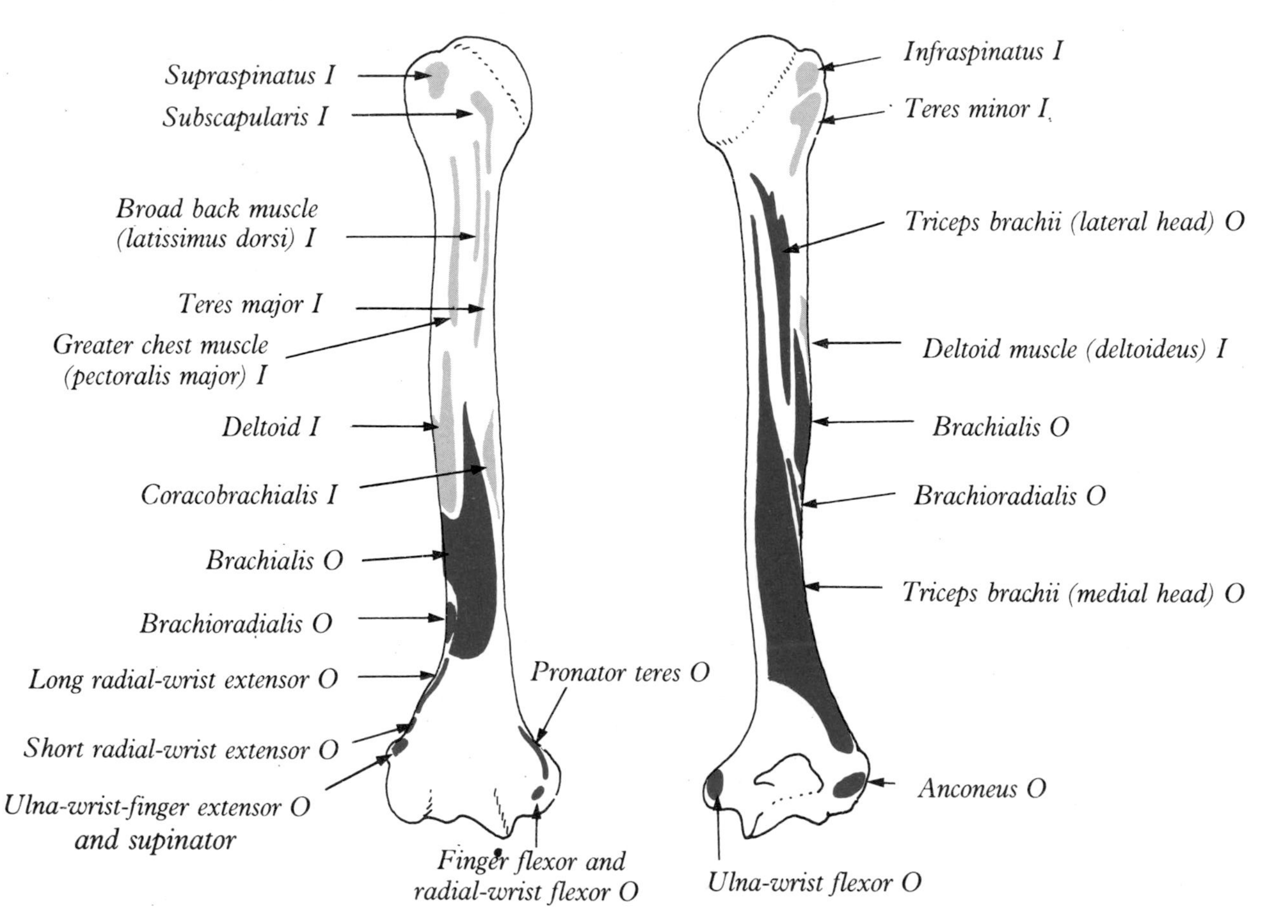

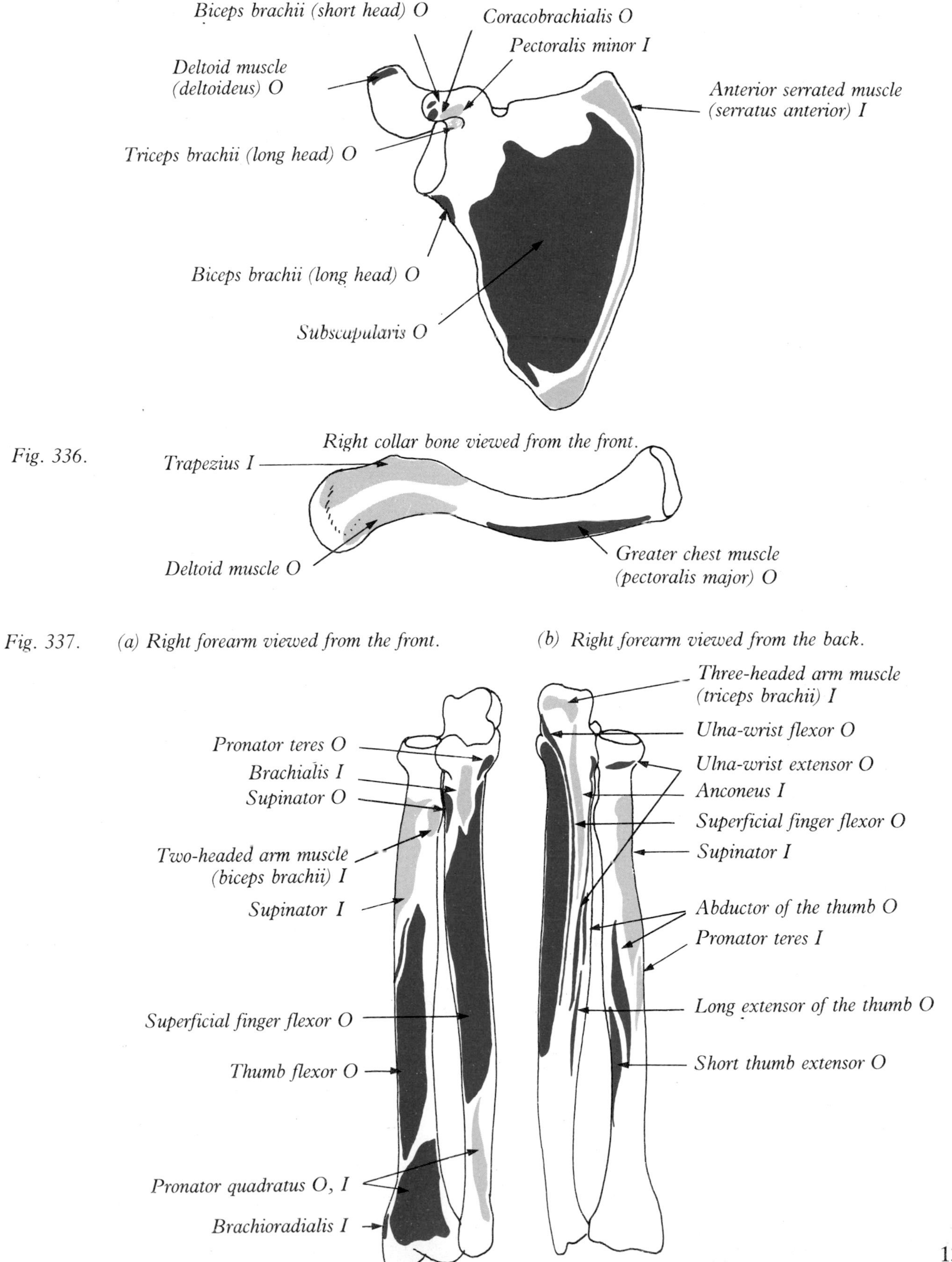

Fig. 335 *Right shoulder blade viewed from the front.*

Fig. 336. *Right collar bone viewed from the front.*

Fig. 337. *(a) Right forearm viewed from the front.* *(b) Right forearm viewed from the back.*

REFERENCES

Anderson B. — Stretching ISBN 0-9601066-1-8

Basmajian J.V. — Muscles Alive 4th ed (Illus) 1979
Williams & Wilkins Co, USA ISBN 0-683-00413-1

Biomechanics — International Series on Biomechanics, Volume I-VI
University Park Press, USA ISBN 0-8391-1242-4

Dagg A. — Running Walking and Jumping
Wykeham publications, London 1977 ISBN 0-85109-530-5

Doherty K. — Track and field omnibook ISBN 0-911520-73-2

Dyson G. — The mechanics of athletics
University of London press Ltd 1962 SBN 340 08905 9

Eaves G. — Diving
A.S. Barnes & Co, Cranbury N.J. USA NBN 99-042-1168-X

Ekstrand J. — Soccer injuries and their prevention
Linköping University Medica Dissertations No 130 SBN 91-73-72-526-9

Frankel-Burstein — Orthopaedic biomechanics
Lea & Febiger, Philadelphia 1970 ISBN 8121-0090-5

Harris R. — Kinesiology
Houghton Mifflin Co, Boston USA 1977 ISBN 0-395-20668-5

Hay J. — The biomechanics of sport techniques
Prentice-Hall INC. N.J. USA ISBN 0-13-035-139-3

Hjortsjö C-H. — Rörelseapparaten Spec. del.
Gleerups, Lund 1967

Hochmuth — Biomechanik sportlicher Bewegungen
Sportverlag, Berlin 1974 Lic No 140355-5-74-9004

Janda V. — Muskelfunktionsdiagnostik
Studentlitteratur, Lund 1976 ISBN 91-44-11351-X

Liljedahl-Gillqvist — Idrottsskador
Liber 1974 ISBN 91-47-01109-2

Petersson, Renström — Skador inom idrotten
Sv. Riksidrottsförbundet 1977 ISBN 098 2034-X1

Styrka och Rörlighet — Sv. Riksidrottsförbundet/Utb. NBN 99-031-6079-8

Sölveborn S. — Boken om stretching
Sölve Bok, Ystad 1982 ISBN 91-970392-0-9

Thompson C.W. — Manual of Structural Kinesology 1977
C.V. Mosby Co, St. Louis USA ISBN 0-8016-4939-0

Tittel K. — Beschreibende und funktionelle Anatomie des Menschen
Gustav Fischer Verlag, Stuttgart

Tveit P. — Bevegelseslaere
Universitetsforlaget, Norge ISBN 82-00-25923-4

Williams-Lissner — Biomechanics of Human Motion SBN 0-7216-9440-3

Williams-Sperryn — Sports medicine ISBN 0-7131-4275-8

Åstrand-Rodahl — Textbook of work physiology McGraw-Hill 1977 ISBN 07-002405-7